Spaceship Earth
How Long Before We Crash?

James Powell

With Jesse Powell and James Jordan

Graphic Illustrations by Douglas Rike

Cover Design by Aspen Larsen

Also by Gordon Danby, James Powell with James Jordan
> *The Fight for Maglev, Making America the World Leader in 21st Century Transport*
>
> *Maglev America, How Maglev Will Transform the World Economy*

By James Powell, George Maise, and Charles Pelligrino
> *StarTram, The New Race to Space*

By James Powell, Jesse Powell and James Jordan
> *Silent Earth Will Humans Give Up Fossil Fuels?*

By James Powell, Gordon Danby, with Robert Coullahan, Ernest Fazio, Fletcher Griffis, George Maise, Charles Pellegrino, Jesse Powell, John Powell, John Rather and James Jordan, Managing Editor
> *7 Big Projects for a Better World*

Spaceship Earth
How Long Before We Crash?

Copyright 2018 James Powell

All rights reserved

For permission to reproduce or transmit in any form or by any means the contents of or any part of this book, contact: james.jordan@cox.net

ISBN-13: 978-1986077736

ISBN-10: 198607773X

Library of Congress Control Number: 2018903157

Printed in the United States

For information about bulk purchases, please contact:

james.jordan@magneticglide.com

Dedication

We dedicate Spaceship Earth to all the humans working to preserve Earth's natural beauty, and who oppose those exploiting and destroying the planet for personal gain. We need to pass on to our children and grandchildren a healthy, beautiful, livable World, not a World of deserts, starvation, immense storms, Billions of people displaced by rising sea levels, plagues, and deadly heatwaves, with most of Earth's species gone in the 6th Mass Extinction.

Quotations Describing Spaceship Earth

Henry George, Progress and Poverty (1879). "It is a well-provisioned ship, this on which we sail through space. If the bread and beef above decks seem to grow scarce, we but open a hatch and there is a new supply, of which before we never dreamed."

George Orwell, The Road to Wigan Pier (1937), "The world is a raft sailing through space with, potentially, plenty of provisions for everybody; the idea that we must all cooperate and see to it that everyone does his fair share of the work and gets his fair share of the provisions seems so blatantly obvious that one would say that no one could possibly fail to accept it unless he had some corrupt motive for clinging to the present system."

Adlai Stevenson, Speech to the UN (1965) "We travel together, passengers on a little space ship, dependent on its vulnerable reserves of air and soil; all committed for our safety to its security and peace; preserved from annihilation only by the care, the work, and, I will say, the love we give our fragile craft."

Buckminster Fuller, Operating Manual for Spaceship Earth, (1968) "…we can make all of humanity successful through science's world-engulfing industrial evolution provided that we are not so foolish as to continue to exhaust in a split second of astronomical history the orderly energy savings of billions of years' energy conservation aboard our Spaceship Earth."

UN Secretary U-Thant, Earth Day (1971), "May there only be peaceful and cheerful Earth Days to come for our beautiful Spaceship Earth as it continues to spin and circle in frigid space with its warm and fragile cargo of animate life."

Walt Disney World Epcot Park. Spaceship Earth, a 180-foot diameter geodesic sphere

Table of Contents

Dedication iii

Quotations Describing Spaceship Earth iv

Table of Contents v

The Purse Seine vi

Foreword vii

Preface viii

Acknowledgement x

God's Grandeur xi

    Chapter 1 1

    Chapter 2 9

    Chapter 3 17

    Chapter 4 77

    Chapter 5 93

    Chapter 6 105

    Chapter 7 131

    Appendix A 137

    Appendix B 163

About the Authors 179

The Purse Seine

*Our sardine fishermen work at night in the dark
of the moon; daylight or moonlight
They could not tell where to spread the net,
unable to see the phosphorescence of the
shoals of fish.
They work northward from Monterey, coasting
Santa Cruz; off New Year's Point or off
Pigeon Point
The look-out man will see some lakes of milk-color
light on the sea's night-purple; he points,
and the helmsman
Turns the dark prow, the motorboat circles the
gleaming shoal and drifts out her seine-net.
They close the circle
And purse the bottom of the net, then with great
labor haul it in.*

*I cannot tell you
How beautiful the scene is, and a little terrible,
then, when the crowded fish
Know they are caught, and wildly beat from one wall
to the other of their closing destiny the
phosphorescent
Water to a pool of flame, each beautiful slender body
sheeted with flame, like a live rocket*

*Lately I was looking from a night mountain-top
On a wide city, the colored splendor, galaxies of light:
how could I help but recall the seine-net
Gathering the luminous fish? I cannot tell you how
beautiful the city appeared, and a little terrible.
I thought, We have geared the machines and locked all together
into inter-dependence; we have built the great cities; now
There is no escape. We have gathered vast populations incapable
of free survival, insulated
From the strong earth, each person in himself helpless, on all
dependent. The circle is closed, and the net
Is being hauled in. They hardly feel the cords drawing, yet
they shine already. The inevitable mass-disasters
Will not come in our time nor in our children's, but we
and our children
Must watch the net draw narrower,*

Robinson Jeffers, The Purse Seine (Abridged)

Foreword

The net is drawing tighter for humans. Civilization is becoming more and more complex and increasingly fragile. In ancient times, there were separate civilizations. Sometimes, they fought each other, sometimes, they lived in peace. Sometimes, a civilization would collapse, but the other civilizations survived. The Mayan civilization collapsed because of their slash and burn agriculture, but the Aztec and Inca civilizations survived, at least until the Spanish conquistadors invaded them.

And Earth survived. Yes, there were local environmental catastrophes – the Mayas, the Anasazi, the Easter Islanders – but the World, as a whole, went on.

Today is different. We have a global, highly interactive civilization. What one country does affects the rest of the World, as evidenced by the oncoming environmental catastrophe from fossil fuels. To avert catastrophe to all of humanity, we must act together to transition to non-fossil energy as soon as possible. Nations and individuals must begin to act in the interest of all of humanity, not just their personal interest.

Preface

70,000 years ago, 7,000 humans roamed the plains of Africa, looking for plants and animals to feed themselves, hoping they would not be food for some other species.

Today, there are 7 Billion humans, a million times more than the human population 70,000 years ago, spread over all the world's continents. No longer just hunter-gatherers, humans grow their own food, have large, comfortable and safe homes, travel by cars, planes, and ships everywhere in the world, visit and explore space, communicate by phones, computers, radio, TV, movies, letters, books and light power cities and homes with electricity and fuels.

Humanity is no longer just one of the many species on Spaceship Earth, but now completely dominates it. We kill other species, cut down vast forests, divert and dam big rivers, mine massive quantities of fuels and minerals, pollute lakes, rivers, oceans, the atmosphere, and cause the planet's temperature to increase and sea levels to rise.

Humanity has been successful. While we have had many wars and conflicts throughout our history, killing and hurting countless numbers of our fellow humans, life has become better for humanity. We want and expect life to get even better in the future. Fewer wars and conflicts, higher standard of living for everyone, new worlds to explore. No end to growth.

Unfortunately, this view of our future is not going to happen if we continue on humanity's present path. Modern society depends on massive consumption of fossil fuels. Every year, if coal, oil, and natural gas are burned for energy to sustain modern society, 30 Billion tonnes of carbon dioxide are emitted into Spaceship Earth's atmosphere, warming the planet. On our present fossil fuel path, the carbon dioxide concentration in the atmosphere will be 1400 parts per million by 2100, 5 times higher than the 280 parts per million concentration in 1800, before we began massively consuming fossil fuels.

Spaceship Earth cannot withstand the massive environmental destruction that will result from such large increases in atmospheric carbon dioxide concentration. Droughts and severe storms will increase, there will be massive deforestation and decreases in crop yields, deadly heat waves, species extinctions, plagues, rising sea levels and coastal flooding, ocean acidification and extinction of much of marine life, etc., etc.

The above consequences of global warming would wipe out modern society, and possibly the human species. Spaceship Earth would crash. Even if the human species did not go extinct, we would forever remain in a primitive state, with significant access to fossil fuels not possible.

To avoid the crash of Spaceship Earth, humanity must soon, in the next few years, begin to:

- Transition from fossil fuels to low cost, clean, sustainable renewable energy services that are not limited in supply.
- Extract massive amounts of carbon dioxide from the atmosphere at acceptable cost and store it in geologically stable underground formations.
- Ensure sustainable soil and water resources that can reliably supply the food and water that modern society requires.

Our book describes 3 new technologies that can meet the above objectives and avoid having Spaceship Earth crash.

To transition from fossil fuels, Maglev the new technology of magnetically levitated transport, can magnetically launch space solar power satellite systems into orbit at very low cost, enabling the satellite systems to beam unlimited amounts of electric power world-wide at much less cost than present electric sources.

To extract massive amounts of carbon dioxide from the atmospheric, air flow can be directed through beds of small particles that absorb the carbon dioxide at ambient temperature. The carbon dioxide is then desorbed by modestly heating the particles, with the collected carbon dioxide then pumped into underground rock formations where it reacts with the rock to form geologically stable material.

To provide the sustainable water supplies needed for long-term food production, the new technology of very low-cost desalination of seawater using ocean thermal energy conversion (OTEC) plants located on floating large, thermally insulated ice structures is described. One OTEC-ICE Plantship with a capital cost of less than 1 billion dollars can produce 2 Billion gallons per day of fresh water, at a cost of only 10 cents per 1,000 gallons. OTEC-ICE Plantships can also produce very large amounts of electric power, thousands of megawatts at very low cost, on the order of 2 cents per kilowatt hour (electric)

Spaceship Earth describes the 3 new technologies in detail and projects their capital and operating costs.

Implementation of these 3 new technologies will avert the crash of Spaceship Earth.

Acknowledgement

The authors wish to express their gratitude and thanks to our following colleagues who have helped to develop the new technology concepts described in Spaceship Earth.

They include Bob Coullahan, Bud Griffis, George Maise, Charlie Pellegrino, John Powell, John Rather, Douglas Rike, Barbara Roland, John Skaritka, Meyer Steinberg, and Tom Wagner.

All of us want a better World, and we hope that the technologies described here will help that goal.

About the Cover

A special acknowledgement of Aspen Larsen, the artist who captured the intention of the authors in her vivid depiction of images that inspired the authors of Spaceship Earth.

Evolving from a simple pencil sketch by James Powell of Noah's Ark with dead animals surrounding the ark, it inspired the selection of NASA images of space as a background for the concept of space solar satellites in geosynchronous orbit collecting energy from the Sun and then converting the energy to low energy microwaves for beaming to the Earth and providing he World with a very cheap and reliable source of non-fossil energy.

Aspen Larsen selected an image of the Earth that was half in sunlight and half dark with cities lighted with electricity. Noah's Ark was at first part of the front cover but in the well-known principle of keeping the art simple, the 1846 folk art painting of Noah's Ark by American Edward Hicks was moved to the back cover as the foundation of the short description of what Spaceship Earth is about: affordable electricity, desalination of ocean water, and a system for capturing and sequestering massive amounts of carbon dioxide from the atmosphere.

God's Grandeur

By Gerard Manley Hopkins

The world is charged with the grandeur of God.
   It will flame out, like shining from shook foil;
   It gathers to a greatness, like the ooze of oil
Crushed. Why do men then now not reck his rod?
Generations have trod, have trod, have trod;
   And all is seared with trade; bleared, smeared with toil;
   And wears man's smudge and shares man's smell: the soil
Is bare now, nor can foot feel, being shod.

And for all this, nature is never spent;
   There lives the dearest freshness deep down things;
And though the last lights off the black West went
   Oh, morning, at the brown brink eastward, springs —
Because the Holy Ghost over the bent
   World broods with warm breast and with ah! bright wings.

Source: Gerard Manley Hopkins:
Poems and Prose (Penguin Classics, 1985)

Chapter 1

Life on Spaceship Earth

Most humans believe that they live on a vast, unlimited Earth that will support an ever-increasing human population with an ever better quality of life. As the years go on, we will have more and tastier food, purer water and air, bigger houses, cheaper medical care, more travel and better entertainment, etc., etc. Today, there are 7 Billion humans in the world, by 2050 there will be 9 Billion. By 2100, 11 Billion.

Sorry, but it won't happen. Not only is continued growth of population and the unlimited use of Earth's finite resources not possible, but if we keep on our path Spaceship Earth will crash in the coming decades, wiping out humanity.

It's not hard to understand why humanity doesn't want to believe that Spaceship Earth will soon crash. Humans are inherent optimists. We like to think that we will overcome obstacles and things will get better. Humans have to be optimistic to survive. Not just us, but all life forms are optimists – that's Darwinism.

Human history reinforces our inherent optimism. 70,000 years ago, things looked pretty bad for homo sapiens. As a species, we emerged about 200,000 years ago in Africa, along with a number of other hominoid species – Neanderthals, homo erectus, etc. 70,000 years ago, the total homo sapiens population was about 7,000 individuals, living off the plants and animals growing in Africa.(1)

Then the world changed. The Toba Volcano in Sumatra, Indonesia, erupted, spewing 2800 cubic kilometers (670 cubic miles) of vaporized rock into the atmosphere.(2) This was 1,000 times more vaporized rock than Vesuvius expelled in 79 AD.

To picture the volume, Figure 1.1 shows a map of Long Island in New York State, 1400 square miles in surface area. The Toba volcano hurled into the atmosphere a volume of rock equal to a 1/2-mile-thick layer of area equal to the surface of Long Island.

FIGURE 1.1. MAP OF LONG ISLAND, NEW YORK STATE

Toba's effects were felt worldwide. 2 inches of ash were deposited over South Asia, the Indian Ocean, and the South China Sea. Global temperatures dropped as much as 20 degrees, due to the decline in solar radiation. Plant and animal life in Africa and other parts of the World shrank, with the "nuclear winter" lasting for many years.

Following the eruption, a portion of the 7,000 homo sapiens living in Africa migrated to the other continents, looking for food. Figure 1.2 illustrates the routes they followed. They spread out into the Neanderthal territory in Europe area, the homo erectus territory in Asia, the Mideast, and the rest of Africa.

FIGURE 1.2 MAP OF EARLY HUMAN MIGRATIONS

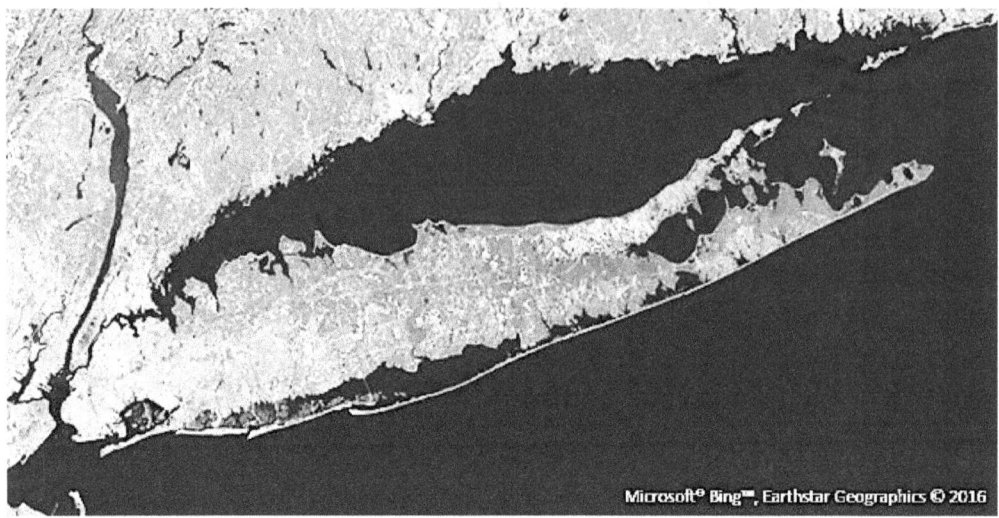

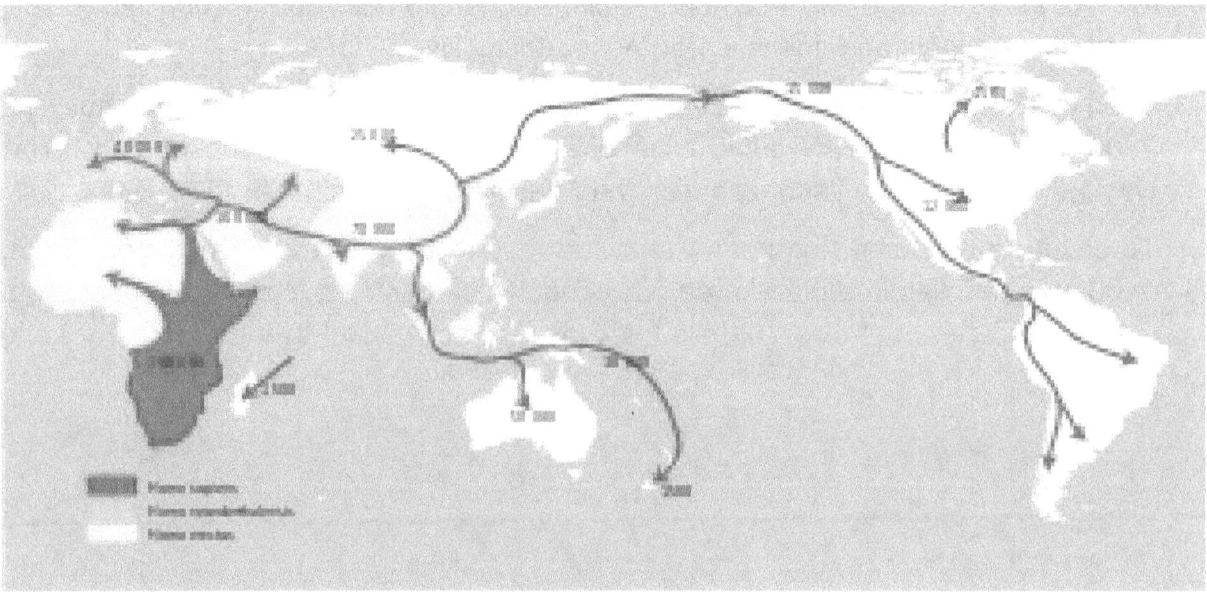

1. Homo sapiens, 2. Neanderthals, 3. Early hominins

They did not stop there. Their journeys continued on to North and South America, Australia and lots of islands only skipping the trip to Antarctica.

Pretty good for a World total of only 7,000 homo sapiens 70,000 years ago. Today, 21,000 people work in the Empire State Building (3) on Manhattan Island in New York City.(Figure 1.3), three times humanity's world population 70,000 years ago.

FIGURE 1.3   **FIGURE 1.4 MAP OF AFRICA**

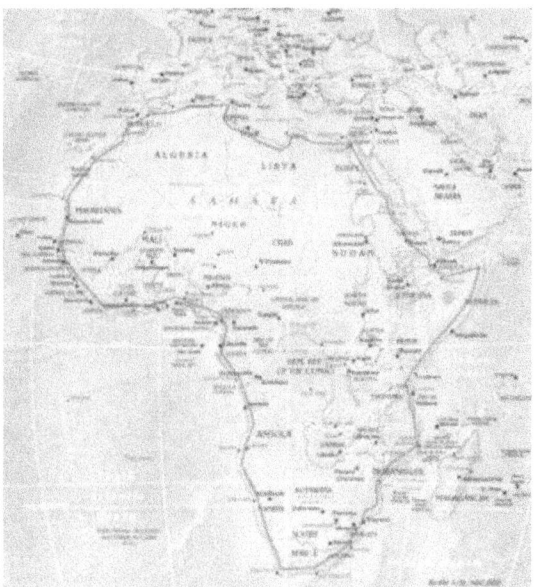

Picture how spread out the 7,000 humans were. Africa (Figure 1.4) has a land area of 11.7 million square miles. Human population density in Africa 70,000 years ago was 1700 square miles per person. Assuming the 7,000 humans were spread out evenly across Africa, one would have to walk 41 miles to meet one's nearest neighbor.

We've grown a lot since then. Today, world population is 7 Billion people, not 7 thousand, an increase by a factor of 1 million. Figure 1.5 shows a graph of world population as a function of time since 10,000 BC.

FIGURE 1.5 GRAPH OF WORLD POPULATION SINCE 10,000 BC

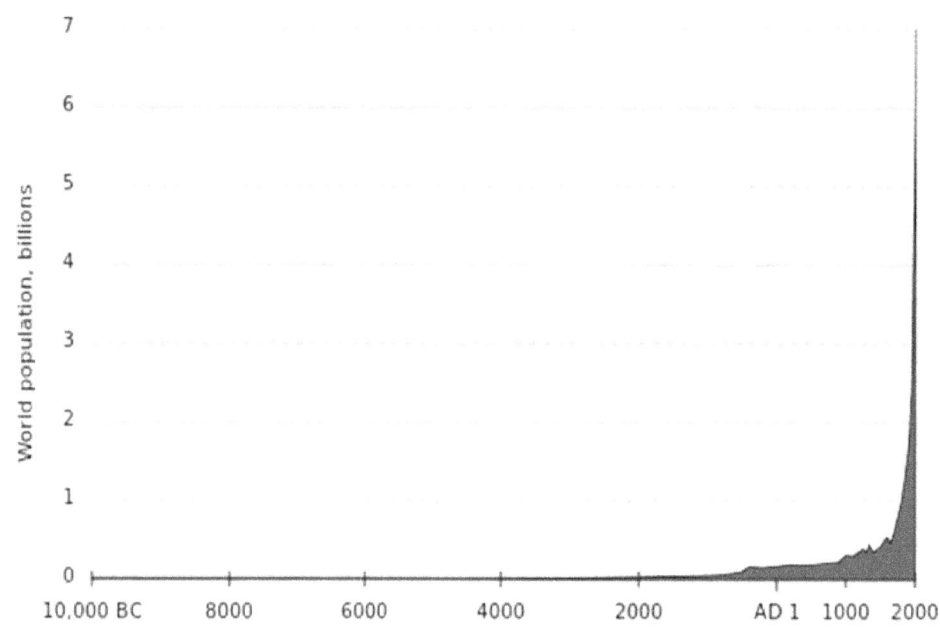

Put into numbers and years, world human population as a function of time is (4):

Year	World Human Population
10,000 BC	4 million
1,000 BC	50 million
1 AD	200 million
1,000 AD	400 million
1500 AD	458 million
1750 AD	791 million
1800 AD	1 Billion
1900 AD	1.65 Billion
1960 AD	3 Billion
1975 AD	4 Billion
1990 AD	5.3 Billion
2000 AD	6.1 Billion
2010 AD	7 Billion

The Agricultural Revolution in 10,000 BC enabled substantial population growth and much more complex societies, starting through the Sumerian, Ancient Egypt, Athens and the other Grecian City States, the Roman Empire, the Ottoman Empire, etc.

However, world population really took off in the 1800s at the start of the Industrial Revolution with the fossil energy from coal, oil and gas that fueled it, became widely available. Without fossil fuels, we would still be an agricultural based society with a population of about 1 Billion.

Now we number 7 Billion humans. Spread out uniformly over the 58 million square miles of Earth's land surface, that's 120 persons per square mile. One doesn't have to walk 41 miles to one's nearest neighbor, as was the case 70,000 years ago in Africa. Today, if humans were spread out uniformly across the Earth, one only has to walk 480 feet to the nearest neighbor.

Even better, one doesn't even need to walk over to the nearest neighbor, with the billions of cellphones and computers in the world. We humans are in constant contact with each other, morning, noon, and night through Facebook, Twitter, and the other social media.

Is humanity willing to give up fossil fuels and the high standard of living that they enable and go back to a lower standard of living, shorter life spans, a horse & wagon type of existence?

Most emphatically, NO. Humans will continue to burn vast amounts of fossil fuels, currently at a world total of 15 Billion tons per year, with carbon dioxide emissions of 36 Billion tons per year.

Humans will only give up fossil fuels and switch to clean energy sources if the new energy sources continue to sustain our life style and are affordable.

Even if new technologies can be implemented that can take the place of fossil fuels while still maintaining our present life style, there will be tremendous political and economic resistance to action from corporations, vested financial interests, politicians and a large fraction of the population.

This is to be expected. Big changes alarm most people – they don't like disruption. They do what is necessary, hope that things will get better, or, like the French Aristocracy, shrug off the needed change, saying "After me the Deluge".

They feel reinforced in their denials about the predictions of a disastrous future by looking back at human history. Over thousands of years, humanity has survived the collapse of many of its societies because of environmental disasters, wars, etc., but has continued to grow and prosper. In the words of Spock, humanity has "lived long and prospered".

However, the historical collapses have generally taken place on small pieces of Spaceship Earth and not seriously damaged Spaceship Earth itself.

Today, Spaceship Earth is in danger of total collapse, as described in the following chapters. If it collapses, humanity will not recover as it previously has; either become extinct or a very small population, maybe 7,000 humans.

References:

1. Recent African Origins of Modern Humans, https://en.wikipedia.org/wiki/Category:Recent_African_origin_of_modern_humans

2. Toba Catastrophic Theory, https://en.wikipedia.org/wiki/Toba_catastrophe_theory

3. Empire State Building, https://en.wikipedia.org/wiki/Empire_State_Building

4. https://en.wikipedia.org/wiki/world_population

Figure Credits:

Figure 1.1, Map of Long Island, NY, Public Domain

Figure 1.2, File: Spreading Homo Sapiens. la.org, https://commons.wikimedia.org/wiki/File:Spreading_homo_sapiens_la.org, author, Nord_Mord_West

Figure 1.3, Photo, Empire State Bldg., Public Domain

Figure 1.4, Map of Pan African Maglev System, Authors, Powell and Jordan

Figure 1.5 Graph of World Population Since 10,000 BC, https://upload.wikimedia.org/wikipedia/commons/b/b7/Population_curve.svg, author, BIT

"The ideas of economists and political philosophers, both when they are right and when they are wrong, are more powerful than is commonly understood. Indeed, the world is ruled by little else. Practical men, who believe themselves to be quite exempt from any intellectual influence, are usually the slaves of some defunct economist. Madmen in authority, who hear voices in the air, are distilling their frenzy from some academic scribbler of a few years back. I am sure that the power of vested interests is vastly exaggerated compared with the gradual encroachment of ideas. Not, indeed, immediately, but after a certain interval; for in the field of economic and political philosophy there are not many who are influenced by new theories after they are twenty-five or thirty years of age, so that the ideas which civil servants and politicians and even agitators apply to current events are not likely to be the newest. But, soon or late, it is ideas, not vested interests, which are dangerous for good or evil."

The conclusion of John Maynard Keynes in The General Theory of Employment, Interest and Money, published, February, 1936

The Ship Titanic

They built the ship Titanic
To sail the ocean blue,
They thought they had a ship that the water
Would never go through,
But the Lord's Almighty Hand
Knew the ship would never land.
It was sad when the great ship went down...
(Chorus) It was sad, SO SAD! It was sad, SO VERY SAD!
It was sad when the great ship went down
Husbands and wives, little children lost their lives!
It was sad when the great ship went down

When the ship left England for the shore
The rich refused to associate with the poor
So they put them down below
Where they were the first to go
(Chorus)

Chapter 2

Critical Actions Needed to Save Spaceship Earth

Spaceship Earth is unsinkable – the belief of most humans, just like the belief of the 2,224 passengers and 892 crew on RMS Titanic, as she sailed out of Southampton, England for New York City on April 10, 1912.(1)

The Titanic had 3 classes of passengers, First, Second, and Third. The rich traveled First Class, with a ticket costing as much as $120,000 in 2015 dollars for a parlor suite and small private promenade deck.(2)

Emigrants to America traveled Third Class, at a ticket cost of $1,000 in 2015 dollars, with children's tickets at approximately $300 in 2015 dollars. Figure 2.1 shows a photo of a Third-Class cabin on the Titanic.

FIGURE 2.1

Inequality has always been with us throughout history. Typically, society does not revolt, as long as the inequality does not get too bad. If the society is growing and conditions for the lower classes are getting better, they put up with their lower standard of living compared to the rich. As long as they have bread, they accept royalty eating cake.

However, if growth slows and stops, the masses lose hope and revolt, storming the Bastille or the Czar's palace.

Today, inequality is a major problem in most of the world, only sustained by the continued growth of the world's Gross Domestic Product.

Current Organization for Economic Cooperation and Development (OECD) projection for World Gross Domestic Product are:

World GDP (Trillions of Dollars)	Date
71	2016
146	2040
182	2050

Because of the very serious problem for Spaceship Earth described later: it appears improbable that these GDP projections will be achieved. A factor of almost 3 growth in World GDP in only 35 years, given our problems with energy, global warming, food and water?

In the decades ahead, the growing inequality between rich and poor in countries will increase the anger of the lower classes, fostering revolutions inside countries and wars between countries.

If inequality, revolutions and wars were the only problems we would face in the coming decades, Spaceship Earth and humanity would continue to survive, as we have done throughout history. We would "live in interesting times" — not pleasant ones — as the saying goes, but massive environmental catastrophe and extinctions would not occur.

However, unless spaceship Earth is not going to crash, the following critical actions must be taken:

Stop using fossil fuels and transition to clean, renewable, and sustainable energy sources that do not damage and degrade the environment.

Remove carbon dioxide from the atmosphere to prevent runaway global warming and bury it in a geologically stable form.

Ensure sustainable soil and water resources that meet the food and water needs of the World population.

The first critical action, the use of fossil fuels.

Present human society is completely dependent on consumption of incredible amounts of energy and cannot survive without it. Currently, each of us 7 billion humans consume on average the equivalent of 3,000 watts of primary energy to generate electrical power, heat our homes and businesses, fuel our autos, trucks, and airplanes, manufacture our TV Sets, cellphones, air conditioners, grow our food, provide shelter, clothing, healthcare, etc., etc.

The energy input dwarfs the energy we consume as food to stay alive. Hunter-Gatherers in ancient Africa living on 2,500 calories per day, consumed energy at a rate of approximately 100 watts, a factor of 30 less than the average human now consumes daily.

And that's at current consumption rates. In developed countries, e.g. the US, Europe, Japan, Canada, etc., the average energy consumption per person is much greater than the world average. As the world population and living standard increases in the coming decades, both total energy consumption and energy production will increase substantially.

If we keep consuming fossil fuels and increasing energy consumption, devastating environmental effects will occur. By 2100, with the lifetime of the children now being born, coastal cities will be flooded all over the world, hundreds of millions of people will be displaced from where they now live, global temperatures will be much greater, Arctic sea ice and the world's glaciers will be gone, millions will die every year from heat waves, the oceans will become so acidic that a large fraction of the marine species will go extinct, large sections of the Greenland and Antarctic ice sheets will have melted away, contributing to rising sea levels, severe hurricanes and storms will be much more frequent and stronger, droughts and increasing crop failures, wildfires, etc., etc.

The present global actions and plans to transition from fossil fuels to clean renewable energy sources are completely inadequate and will not avoid environmental catastrophe.

Implementation of beamed power from space solar satellites down to Earth offers a path to fully meet humanity's needs for clean, environmentally safe energy without fossil fuels, and with major economic benefits from its much lower cost than those of present energy sources.

Chapters 3 and 4 describe how space power satellite systems can be developed and implemented. Chapter 3 describes the StarTram launch system, based on magnetic levitation and acceleration of spacecraft to orbital speeds, enabling them to be placed into orbit at much lower cost than using rockets. Using StarTram launch, space power satellite systems could be placed into GEO (geosynchronous orbit) for 1/100th the cost of orbiting them using rockets.

Chapter 4 describes the technology of beaming power down to Earth from Space Solar Satellites, projected cost, and a potential schedule for its implementation.

The 2nd critical action is to remove carbon dioxide from the atmosphere.

It is not generally realized, even by those actively working to reduce the greenhouse gas emissions from fossil fuels and their gas emissions from fossil fuels and their effect on climate that even if successful, Earth may still be headed for environmental disaster.

Greenhouse emissions and global warming will not stop, even if humanity completely stops using fossil fuels and transitions to clean renewable energy sources.

There are enormous amounts of organic carbon locked in the frozen permafrost of Canada and Siberia. As the permafrost thaws and the organic carbon oxidizes, large amounts of carbon dioxide will be released into the atmosphere.

In addition, there also are enormous amounts of marginally stable methane hydrates material on the ocean sea beds. As the oceans warm, the methane hydrates are already starting to decompose, releasing methane into the atmosphere.

Methane is a much more potent greenhouse gas than carbon dioxide – 20 to 30 times more potent. If substantial amounts of the marginally stable methane hydrates decompose, there will be a massive global catastrophe, comparable to the Permian extinction 250 million years ago, which wiped out more than 90% of the Earth's species.

At some point as the Earth continues to warm and the greenhouse gas emissions from the heated permafrost and decomposing methane hydrates in the ocean sufficiently increase, global warming will continue and accelerate, even if humans stop consuming fossil fuels.

When will this point occur, termed the "Trigger Point". Nobody knows. We may have already passed it. Global temperature has increased by 1.5 degrees Celsius above pre-industrial levels, and is on track to increase 2.0 degrees Celsius by 2036.

Humanity cannot risk reaching the Trigger Point. In addition to transitioning from fossil fuels to clean energy sources, as described in Chapters 3 and 4, we must begin to extract carbon dioxide from the atmosphere and bury it in long-term geologic formations.

Currently, we emit more than 30 Billion tonnes of carbon dioxide into the atmosphere, with an annual increase in its atmospheric concentration of 2 parts per million. The CO_2 concentration presently at 400 parts per million (ppm), will be more than 600 ppm by 2100 if we continue consuming fossil fuels. We will have gone way beyond the Trigger Point by then.

To ensure that humanity avoids runaway global warming, we must develop and implement technologies to remove massive amounts of carbon dioxide from the atmosphere and geological store it.

As a goal, we should aim at the capability to remove 30 Billion tonnes of CO_2 per year from the atmosphere. As we transition from fossil fuels to clean renewable energy sources, this will help to maintain CO_2 atmospheric concentration at roughly the current level of 400 ppm, hopefully avoiding runaway global warming.

Removing such a large volume of CO_2 from the atmosphere will be an enormous project, the biggest that humanity has ever undertaken. It will be very expensive, and involve massive construction.

Chapter 5 and the Appendix describes a promising technology for CO_2 removal, based on absorption of CO_2 in the atmosphere that flows through beds of small particles, which can be heated to release pure CO_2 for sequestration in stable geologic formations.

The beds of absorber particles can cycle many times, first absorbing carbon diode from the atmosphere at very low concentrations, 400 parts per million, then heating the absorber particles to desorb pure CO_2 at much high pressure. The absorber particle beds are then returned to ambient temperature, with atmospheric air flowing through them to absorb more CO_2 from the atmosphere.

The extracted CO_2 is then pumped at high pressure down into rock formations, where it reacts with the minerals in the rock formation to form geologically stable materials.

It is vital that the world should start aggressive development of technologies that can extract CO_2 from the atmosphere as soon as possible. The extraction technology described in Chapter 5 and the Appendix is not the only potential extraction technology. Other technologies should also be investigated and tested, so that the most effective extraction approach can quickly begin implementation.

Waiting until global warming becomes so severe before humanity recognizes that its effects will be disastrous, risks passing the Trigger Point, runaway global warming, and mass extinction.

The third critical action is to ensure adequate food and water for the world's population.

Chapter 6 describes the present and projected status of the world's soil and water systems, and the actions needed to ensure sustainable supplies of food and water for humanity.

Much of the World's farmlands are experiencing serious problems that affect their capability to grow food.

- Severe droughts
- Depleted aquifers
- Topsoil erosion
- Increasing salinity of topsoil
- Reduced crop yields due to increasing global temperatures
- Reduced crop yields due to increased crop blights and insect populations

As an example of the drought problem, in 2014 92% of California was in the D-3 and D-4 extreme and exceptional drought conditions, with 58% of California in the worst, D-4, condition. Large sections of the Western US are experiencing major droughts, with millions of acres being burned by wildfires. The droughts have reduced crop yields.

To cope with the severe droughts, farmers are over-pumping underground aquifers for the water to grow crops. This is a very serious problem. In many areas, to reach water, one has to drill down hundreds of feet. As the aquifers are exhausted the ground level goes down in some places as much as 30 feet, compacting soil so that it cannot replenish its water if rain does come. At some point, the aquifers will be completely exhausted, and desertification will ensue.

Excessive pumping of the aquifers not only drains them, but it also increases soil salinity, due to salt in the water destroying the ability of farmland to grow crops.

Topsoil erosion is another very serious problem. Many areas have lost much of their topsoil due to flooding from severe storms, poor irrigation, and other reasons. Once topsoil is lost, it takes hundreds of years to replace it.

In addition to the above problems, croplands are suffering from increased attack by blights and insects, which flourish as temperature increases.

Moreover, increasing temperatures also reduce crop yield. As temperature levels rise too far, plants will grow more slowly and beyond a certain temperature will not produce seed. Soybeans will not reproduce at temperatures above 102 degrees Fahrenheit.

With a 4 degrees Centigrade (7.2 degrees Fahrenheit) rise in global temperature by 2100, the average decrease in crop yield for 5 major crops – African maize, Asian rice, Indian wheat, US maize, and US soybeans would be 50%. This would not support the 11 Billion population projected for 2100 AD.

Clearly, if we are to avoid a catastrophe in our food and water supplies, we must begin to develop technologies that can cheaply desalinate water for farmers, prevent soils erosion, and avoid increased soil salinity. Moreover, extracting CO_2 from the atmosphere as described in Chapter 5 and transitioning to renewable energy sources will help to reduce global warming, prevent higher temperatures, and ensure good crop yields.

Key to ensuring sustainable food supplies to humanity in the coming decades will be the development of the capability to desalinate seawater at low cost and distribute the fresh water drought area for growing crops.

Chapter 6 describes a promising new technological approach for desalination using ocean thermal energy conversion (OTEC) plants located on large floating island structures. Fresh water output per plant would be on the order of 2 Billion gallons daily, at a cost of approximately 10 cents per 1,000 gallons – a sufficiently low price for growing crops.

The OTEC desalination plants can be located offshore from drought regions, e.g. California, Africa, the Mideast, etc., with the fresh water product shipped by pipelines to the drought areas.

The OTEC desalination technology can be quickly developed. If implemented, it will help to ensure the capability to grow crops without having to over pump underground aquifers, that destroy topsoil through increasing salinity.

Chapter 2; References:

1. Sinking of the RMS Titanic, https:en.wikipedia.org/wiki/sinking_of_the RMS_Titanic
2. RMS Titanic, https: en.wikipedia.org/wiki/RMS_Titanic

Figure Credits:
Image of Third Class Berth on RMS Titanic – public domain

STARTRAM LAUNCH AND SPACE BASED SOLAR POWER

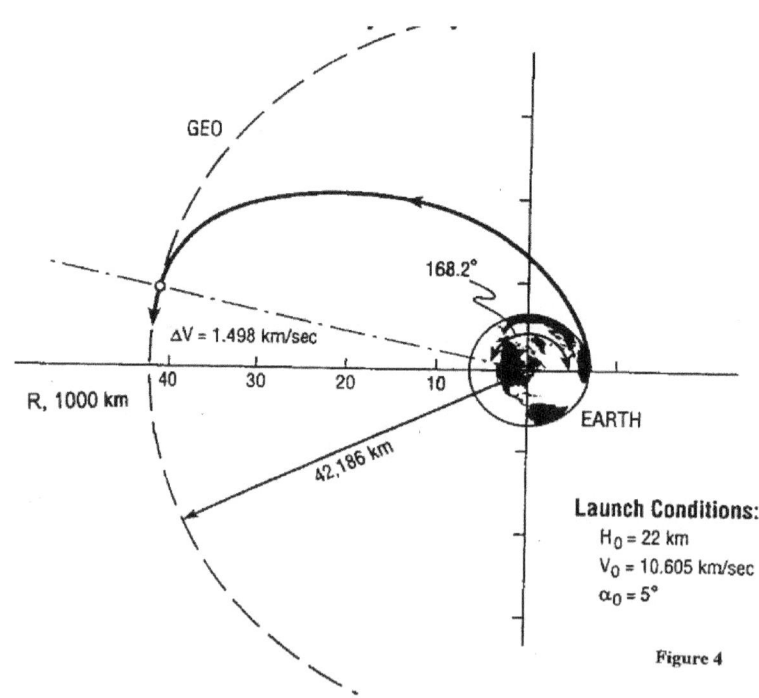

Figure 4

ASCENT TRAJECTORY TO GEOSYNCHRONOUS ORBIT (GEO) USING STARTRAM LAUNCH

Chapter 3

Maglev Launch of Payloads to Orbit at Low Cost

Superconducting magnetic levitation (Maglev), invented by Powell and Danby in 1966, (1) is a fundamentally new mode of transportation. Maglev vehicles have no wheels, no engines, and do not fly. Instead, Maglev vehicles are levitated by the magnetic interaction between the superconducting magnets that they carry and currents induced by ordinary aluminum loops located on the guideway beneath the vehicles.

The levitation is automatic with clearances of several inches between the vehicle and the guideway, and inherently strongly stable. There is no mechanical contact or friction. In the atmosphere, vehicle speed is only limited by air drag, with practical operating speeds of 300 mph. The vehicles are magnetically propelled by the magnetic interaction between their superconducting magnets and electric current that flows in normal conductor loops located on the guideway underneath.

In evacuated tubes without air drag, Maglev vehicles can reach much greater speeds, high enough to be launched into orbit around the Earth. The cost of the electric energy to reach orbital speeds of 8 kilometers per second is only $1 dollar per kilogram at 10 cents per kilowatt hour, a typical utility rate. Adding in the cost of the spacecraft structure and the amortized cost of the launch facility and operations and maintenance (O&M) costs, the total cost of launching 1 kilogram of payload into orbit using Maglev is about $50 per kilogram.

Today, the cost of launching payloads into orbit using rockets is about $5,000 per kilogram, 100 times greater than the projected $50 per kilogram for Maglev launch.

FIGURE 3.1.

PHOTO OF 1ST GENERATION JAPANESE SUPERCONDUCTING MAGLEV OPERATING AT YAMANASHI TEST FACILITY

Superconducting Maglev transport is not a futuristic fantasy. Japan Rail has built and now operates a superconducting Maglev passenger transport system in Yamanashi, Japan. (2) The 30-kilometer-long Maglev guideway (Figure 3.1) reaches vehicle speeds as high as 370 miles per hour, and operates reliably and safely. Cumulative vehicles running distances of hundreds of thousands of kilometers have been achieved with more than 100,000 total passenger trips.

Japan Rail plans to extend the present Yamanashi Maglev system to be a 300-mile route between Tokyo and Osaka, carrying 100,000 passengers daily with a trip time of 1 hour. Over the next couple of decades, Maglev will become a major mode of World transport because of many of its attractive features, lower cost per passenger mile and ton-mile, clean, non-polluting safer and faster travel, and greater energy efficiency than present modes of transport.

Detailed descriptions of Maglev transport are given in 3 books by Powell and colleagues, available at Amazon.com – The Fight for Maglev, Maglev America, and Silent Earth.

Maglev offers more than just high speed Worldwide transport of passengers and freight, however. For example, Maglev can store large amounts of electric energy at low cost from variable wind and solar power sources, to be used during periods of high demand or when the wind isn't blowing or the sun isn't shining.

To store electrical energy, Maglev vehicles move heavy 100 ton blocks uphill from a storage yard at low elevation to a storage yard at a high elevation. The electric energy input required to move the block uphill using magnetic propulsion for the Maglev vehicles can be recovered and fed to the electricity grid by returning the block to the lower storage yard. On the uphill trip, the Maglev propulsion system operates in the storage mode. On the downhill trip, the Maglev propulsion system operates in the generator mode, returning more than 90% of the electric energy used to move the block uphill. Moving a 100-ton block 1000 meters (3,300 feet) stores 250 kilowatt hours of electrical energy.

The most important non-ground transport application, however, will be to launch spacecraft into Earth orbit and beyond at only 1/100th the cost of chemical rockets.

Since the success of the Apollo Program in the 1960's, launching payloads into space has become a vital and major effort for humanity. It enables much better communication, environmental monitoring and weather prediction for Earth's residents, along with exploration of the Solar System and discovery of "what's out there?"

However, the very high cost of launching payloads into space has severely constrained the capabilities for, and political benefits from, operating in space. These constraints result from the high cost of chemical rockets.

To place 1 kilogram of payload into Low Earth Orbit (LEO) costs on the order of $5,000. To Geosynchronous Earth Orbit (GEO) at 33,000 kilometers from Earth, where communication satellites orbit, several times greater. To the Moon, much more, and to Mars, don't ask.

To put space launch costs using chemical rockets in perspective, to fly 1 kilogram of payload from the US to Europe is on the order of $5 per kg, 1/1000th of the cost to launch it into Low Earth Orbit.

Over the years since the Apollo Program, space launch costs have decreased, but only marginally. Many different types of rocket systems have been proposed, developed, tested, operated, and eventually in most cases, died without major reduction in launch cost.

It is not hard to understand why the launch costs using rockets are very high – chemical rockets are very complex structures with many different and expensive components that must work perfectly when the rocket takes off. The component materials operate very close to their physical limits, with the failure of one component likely to cause the whole rocket to fail.

Airplanes land and take off many thousands of times without failure. Rocket failure rate is much greater – on the order of several percent.

Two shuttle disasters, Columbia and Challenger, in a total of 134 flights. Engineers dream about re-usable rockets capable of many flights with very low failure rates, but so far, 45 years after Apollo, efforts continue to develop a realiable system.

A 2nd limitation of chemical rockets is their very low payload fraction. Of the rockets liftoff weight, only several percent is actual payload. More than 90 percent of the take-off weight is propellant and spacecraft structure.

Maglev Launch offers a much lower cost, much more reliable, and much greater payload capability than chemical rocket launch. StarTram, the name for the Maglev Spacelaunch System is described in detail in the book, StarTram: The New Race to Space by James Powell, George Maise, and Charles Pellegrino, available at Amazon.com.

In StarTram, spacecraft are magnetically levitated and accelerated to orbital speed, 8 km/sec or greater, in a long evacuated tube. The spacecraft is equipped with superconducting magnets that interact with the passive aluminum or copper loops imbedded in the wall of the evacuated tube, levitating and stabilizing it as it travels along the launch tube.

The StarTram spacecraft is accelerated by the magnetic interaction of the DC superconducting Magnets on it with pulsed DC currents in aluminum or copper cables imbedded in the launch tube wall. The pulsed DC currents are provided by discharge of superconducting Magnetic Energy Storage (SMES) units located along the launch tube. SMES energy storage is a new energy storage technology that is being implemented in the electric utility industry.

When the StarTram spacecraft reaches its final launch velocity at the end of the launch tube, it exits into the atmosphere through a Magneto Hydro Dynamic (MHD) window that acts to prevent the outside atmosphere from entering the evacuated launch tube.

The StarTram launch tube operates at ground levels in high altitude terrain, with its exit at elevations of several thousand meters above sea level. Atmospheric density at the exit of the launch tube is substantially lower than at sea level, reducing the air drag in the spacecraft as it climbs through the atmosphere to its planned orbital altitude. After reaching orbital altitude, a small rocket motor on the spacecraft fires, establishing the final orbit. The ΔV from the rocket burn is a small fraction, a few percent, of the launch velocity.

The StarTram launch tube exit section is angled, with the StarTram spacecraft to launch at an angle on the order of 10 to 15 degrees, enabling it to ascend up through the atmosphere to orbit. A small portion of its initial launch velocity, several percent, is lost due to air drag as it climbs through the atmosphere. This is compensated for by launching at an initial velocity that is slightly greater than would be required to reach orbital altitude if there were no atmosphere.

The StarTram system described above is the Generation-1 (Gen-1) StarTram. It does not launch humans, only cargo into space. The g forces on the spacecraft are too high, both during the acceleration of the spacecraft in the launch tube and its ascent through the atmosphere to orbit, for humans.

Gen-1 StarTram is near term technology and can be built within the next 10 years, assuming a well-funded, aggressive effort like the Apollo Program and the International Space Station (ISS).

Gen-1 can provide the capability to launch large payloads at very low cost and in much greater amount than chemical rockets can. Depending on design, Gen-1 payload weight would be in the range of 20 to 40 metric tonnes, with a launch cost of approximately $50 per kilogram, slightly more than it costs to fly a kilogram 25,000 miles around the World.

In one year, one Gen-1 StarTram facility that launched 20 metric tonne payloads 10 times a day would launch a total of 70,000 tonnes into orbit, over 500 times the present chemical rocket launch rate of only 100 tonnes annually.

Why is StarTram launch cost so much less than chemical rockets?

First, all the expensive launch equipment remains on the ground, to be used over and over rather than being lost with each launch, the case for chemical rockets. Reusable chemical rockets are proposed as the solution for low cost launch, but even if possible, their cost per launch would be much greater than StarTram, and their lifetime much less. Over 30 years a StarTram facility could launch 100,000 payloads. It is doubtful that a reusable rocket could achieve more than 10 launches.

Second, Gen-1 StarTram does not have to push materials to their operating limits. Its launch equipment on Earth does not operate close to the failure point, but with a large safety margin. Moreover, the equipment can be serviced, maintained, and easily repaired if necessary. Its unit costs will be much less, because it can use existing commercial materials, not special very high-tech ones. Similarly, the StarTram spacecraft structure can be much cheaper than a chemical rocket structure. Its mass fraction is much less and launch cost is low, so more rugged and reliable materials can be used for the spacecraft structure even if they are heavier.

Third, StarTram's spacecraft structure can be much simpler than chemical rockets. No big high thrust rocket engines, only a very small, low thrust engine to finalize its orbit. No second or third stages required, just a simple structure enclosing the payload, with conventional type superconducting cables enclosed in the structure. No long-term refrigeration like those in MRI medical systems, superconducting power transmission and energy storage systems, and high energy particle accelerators. Just cool the cables to their operating temperature and launch.

Fourth, chemical rockets pollute Earth's atmosphere with toxic products from the combustion of their propellants. Even if large scale usage of reusable chemical rockets were feasible, which it is not, their toxic products would destroy Earth's ozone layer, harm human health, and accelerate global warming. StarTram spacecraft uses no propellants and do not harm the atmosphere and human health. Instead, it uses clean, non-polluting electrical energy.

In the long term, it may be possible to evolve the Gen-1 StarTram into the advanced Gen-2 StarTram system that could launch humans into space as well as cargo. The Gen-2 system would have much lower acceleration of the Spacecraft to be compatible with human limitations. The Gen-1 StarTram cargo launch system described later accelerates space craft at 40 g in a 80 kilometer launch tube. A Gen-2 StarTram system with an acceleration of 3 g, acceptable to humans, would be much longer, e.g. 1,000 kilometers in length.

In addition to reducing the g force on passengers in the evacuated launch tube on the ground, by making the tunnel longer, the Gen-2 launch system for human travel to space would have to reduce the g force on the spacecraft as it ascends up though the atmosphere to orbit.

In the Gen-1 system the g force on the spacecraft due to air drag is a maximum as it exits from the launch tube and decreases as it climbs higher into the lower-density atmosphere. Peak deceleration depends on the altitude of the exit from the StarTram Launch tube, being on the order of 10 g at 4,000 meters and 6 g at 8,000 meters.

With optimization of spacecraft design, peak deceleration can probably be reduced to 3 to 4 g, enabling launch of many humans. Also, Gen-2, in which the StarTram Spacecraft leaves the exit from the ground launch tube and enters a magnetically levitated evacuated launch tube (Figure 3.2) that ascends to a higher altitude before it exits into a much lower density atmosphere, with deceleration forces on the order of 2 to 3 g.

The evacuated launch tube would be magnetically levitated by the magnetic interaction between a set of superconducting cables attached to the tube, and a second set of superconducting cables located on the ground beneath. Details of the Gen-2 StarTram System are described in StarTram: The New Race to Space, by James Powell, George Maise, and Charles Pellegrino, available @ amazon.com.

FIGURE 3.2:
VIEW OF STARTRAM EXITING THE LAUNCH TUBE

The Gen-2 StarTram system levitated launch tube is very challenging, involving new technology, and would take much longer to implement than the Gen-1 system. For human launch the best path is probably to initially develop spacecraft with less deceleration due to air drag. This will enable substantial numbers of humans that can experience high g forces to travel into space. As technology developed and magnetically levitated launch tubes become feasible, the levitated tube can be integrated into the system, enabling more humans to be able to launch into space.

How Maglev Launch Works—The Principles

The StarTram spacecraft is magnetically accelerated by the magnetic interaction between a set of DC superconducting (SC) loops attached to the spacecraft and a sequence of "thruster" cables attached to the wall of the evacuated launch tube in which the spacecraft travels.

The thruster cables carry pulsed DC current that interacts with the DC magnetic field produced by the SC cable loops on the spacecraft, generating a magnetic force on the spacecraft, accelerating it in the direction that it is traveling. An equal and opposite magnetic force acts on the thruster cable attached to the launch tube wall, which supports it, and prevents it from moving.

FIGURE 3.3.
THE MAGNETIC LORENTZ FORCE

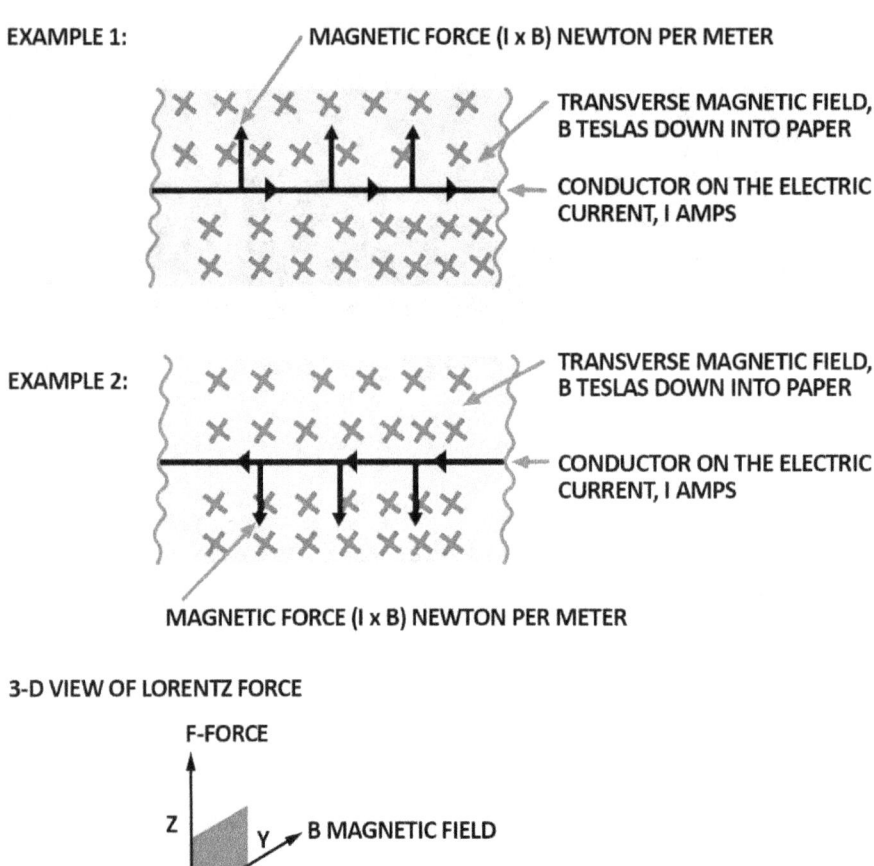

The magnetic force, termed the "Lorentz Force" is illustrated in Figure 3.3. The Force F is given by the vector product of current I and magnetic field B. In SI units, F = I x B Newtons per meter with I in amps and B in Tesla, the force F is given in Newtons per meter of current length. As illustrated in Figure 3.3, with I in the X direction, and B in the Y direction. F is in the Z direction given by the right-hand rule.

As an example of the magnitude of F, a current of 1 million amps in a magnetic field of 1 Tesla produces a force of 1 million Newtons per meter of current length, i.e.

$F = 10^6$ Amps x 1 Tesla = 10^6 Newtons per meter

This represents the maximum force between the current I and the magnetic field B, when the angle between I and B is 90 degrees. If I and B are in the same direction, with zero degrees between them, the magnetic force is zero. If the angle between them is θ degrees, then the force is:

F = I x B sin θ Newtons per meter

The pulsed DC current in the thruster cables comes from superconducting magnetic energy storage (SMES) units located along the evacuated launch tube. Just before the moving StarTram spacecraft reaches a given location, the appropriate SMES unit discharges its stored energy as a pulsed DC current into a sequence of ordinary conductor thruster cables positioned in the wall of the evacuated launch tube.

The pulsed DC thruster cable currents then magnetically interact with the magnetic field produced by the superconducting loops onboard the levitated StarTram spacecraft, accelerating it.

Figure 3.4 illustrates the geometry of a thruster cable in the wall of the evacuated launch tube and the magnetic interaction with the magnetic field produced by the superconducting DC Cable on the StarTram spacecraft.

At distance r from the center of the DC superconducting cable on the spacecraft, the magnetic field strength on the thruster cable is (μ_0 = 4π x 10^{-7} H/m), B_r = ($\mu_0/2\pi$) x I_o/r

At the bottom of the thruster cable, which is at r = 0.2 meters from the center of the StarTram cable with I_o=4 million amps turns (MAT) in the DC Spacecraft SC cable,

B_r (r=0.2) = 2 x 10^{-7} x 4 x 10^6/0.2 = 4 Tesla

At r =1.2 meters, the top end of the 1-meter-long thruster cable,

B_r (r=1.2) = 2 x 10^{-7} x 4 x 10^6/1.2 = 0.67 Tesla

B(r) = $\left(\frac{\mu_0}{2\pi}\right)$ I_o/r = 2 x 10^{-7} I_o/r Tesla

FIGURE 3.4.
THE LORENTZ FORCE AND STARTRAM GEOMETRY

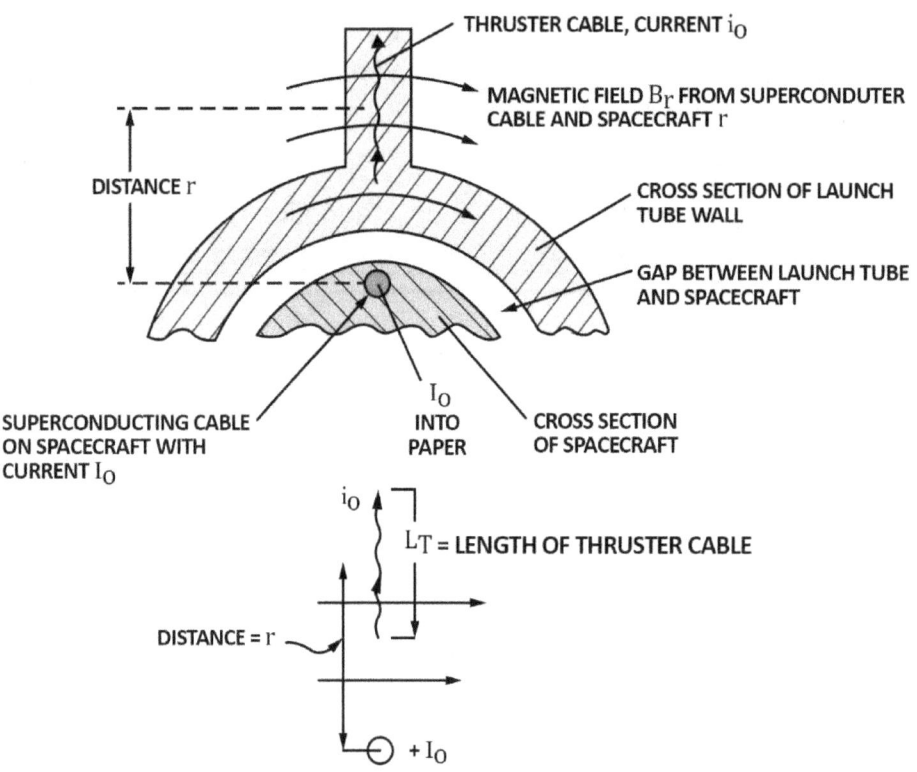

At position r, $B(r) = \left(\frac{\mu_o}{2\pi}\right) Io/r$ Tesla, with Io amp,

Fr = Force on thruster cable at distance r = is $B_{(r)} \times i_o$ Newtons per meter of cable, where i_o, the current in the thruster cable, is in amp,

Ft = Integrated force on thruster cable = $\left(\frac{\mu_o}{2\pi}\right) Io \, i o \int_{r_0}^{r_1} \frac{dr}{r} = \left(\frac{\mu_o}{2\pi}\right) Io \, i o \ln\left(\frac{r_1}{r_0}\right)$

where r_o = distance of base thruster cable from center of superconducting cable and

r_1 = distance of top of thruster cable from center of superconductor cable,

Note: Force on thruster cable is into paper; Force on spacecraft is out of paper

FIGURE 3.5.

MAGNETIC ACCELERATION OF THE STARTRAM SPACECRAFT

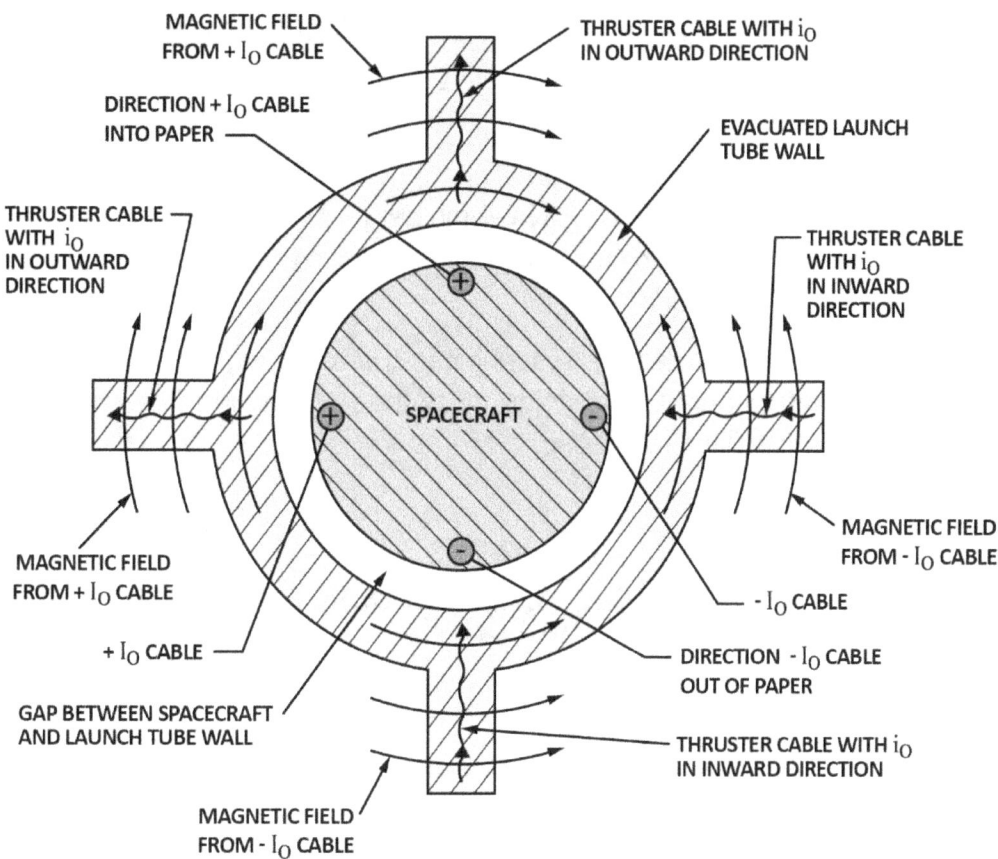

Note: Magnetic force on spacecraft from interaction with thruster cable i_o currents with spacecraft Io currents is directed out of paper. Magnetic forces on thruster cables is into paper.

Integrating F along the 1 meter long vertical thruster cable carrying $i_0 = 10^6$ Amps, the total thrust force on the levitated StarTram spacecraft from the current on the thruster cable is

$F_{TOT} = 2 \times 10^{-7} \times 4 \times 10^6 \times 10^6 \int_{0.2}^{1.2} \frac{dr}{r}$

$F_{TOT} = 0.8 \times 10^6 \ln(6) = 1.43 \times 10^6$ Newtons

With 4 thruster cables in circuits spaced at 90 degree intervals around the evacuated launch tubes (Figure 3.5), each carrying 1 million amps pulsed DC current

$F_{TOT} = 1.43 \times 10^6 \times 4 = 5.7 \times 10^6$ Newtons

Figure 3.6 shows 2 of the 4 thruster cable circuits positioned around the circumference of the evacuated launch tube wall. Each thruster circuit contains a number of thruster cables that are energized with pulsed DC current by the SMES energy storage unit just before the StarTram spacecraft reaches the thruster circuit.

The thruster circuits illustrated in Figure 3.6 have 2 thruster cables that simultaneously magnetically interact with the superconducting DC cables on the spacecraft as it moves along the evacuated launch tube, generating twice as much thrust as a single thruster cable would.

For the example shown, with 1.43×10^6 Newtons per thruster cable, 4 thruster circuits, and 2 active thruster cables per circuit, the total magnetic thrust force on the levitated spacecraft is 11.4×10^6 Newtons

For the StarTram spacecraft design described in the following section, which has a total weight of 28 metric tonnes containing 20 tonnes of payload, the corresponding acceleration is 400 m/sec^2, which is 40g.

= 400 m/sec^2 = 40 g

FIGURE 3.6.
LAYOUT OF THRUSTER CABLE SLAB AND CURRENT WITH STARTRAM SPACECRAFT

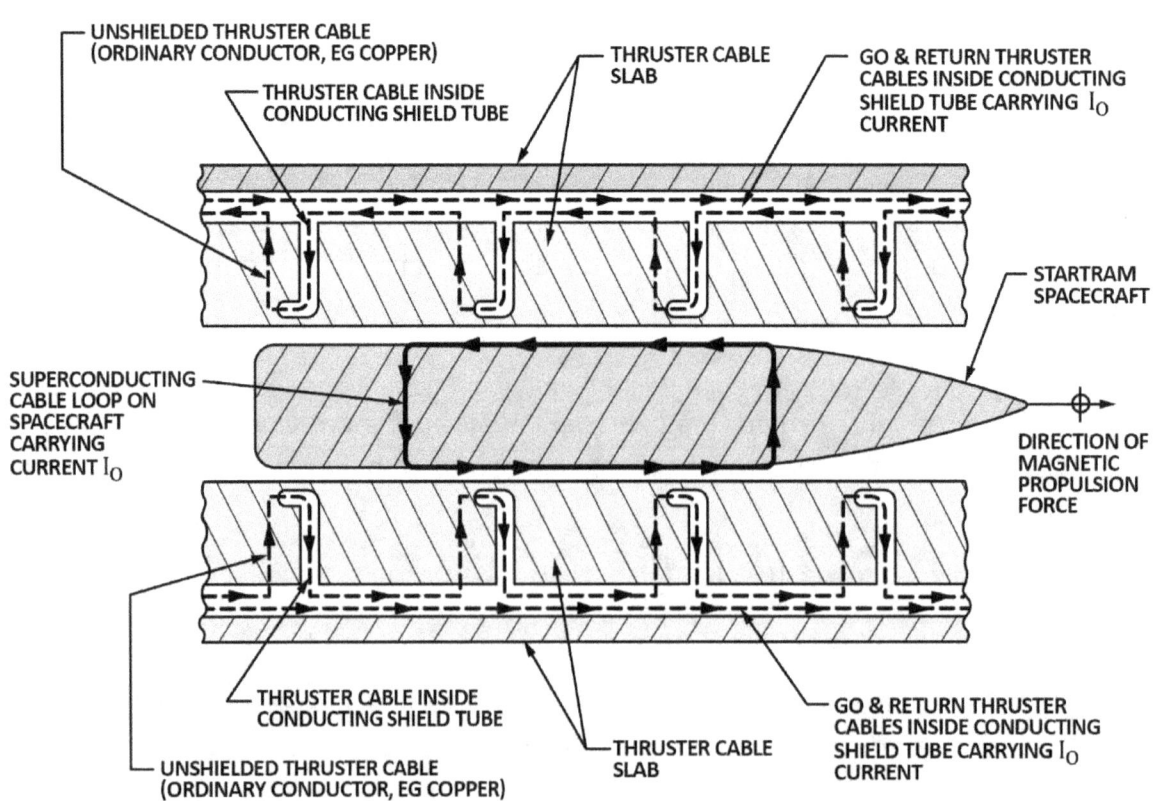

As the spacecraft moves long the evacuated launch tube, the active thruster cable that is left behind the spacecraft is replaced by a new active thruster cable at the front of the spacecraft. There are always active thruster cables producing thrust on the spacecraft, two in each of the four thruster cable circuits that are positioned at 90 degree intervals around the circumference of the launch tube.

The evacuated launch tube is divided into a number of sections, with each section served by its own SMES energy storage unit. As the levitated spacecraft leaves a given section and its thruster cables behind, it enters the next section with new thruster cables that have been energized with pulsed DC current by the SMES unit connected to the section.

The acceleration process continues through the sequence of energized sections, each one further accelerating the spacecraft until it reaches the end of the evacuated launch tube and exits into the atmosphere.

The next part of this report describes the design parameters for a StarTram system that would launch a 28 metric tonne spacecraft containing 20 tonnes of payload at 8 Km/sec using an 80 Km long evacuated launch tube at an acceleration rate of 40g.

The design is illustrative and attractive, but not restrictive. Depending on application, Gen-1 StarTram can launch smaller and lighter spacecraft, or bigger and heavier space craft. The acceleration rate can be greater than 40 g, or less than 40 g, with shorter or longer launch tubes. Launch velocity can be more than 8 Km/sec to reach Geosynchronous Earth Orbit (GEO) and to even higher orbit, or escape Earth's gravity.

Description of Illustrative StarTram Gen-1 Launch System

Table 3.1 gives the parameters for the illustrative StarTram launch system, with Figure 3.7 showing its layout.

TABLE 3.1.

DESIGN PARAMETERS FOR ILLUSTRATIVE GEN-1 STARTRAM LAUNCH SYSTEM

80-kilometer-long evacuated launch tube	2.0-Meter ID Vacuum Tube Wall	900 Gigajoules Kinetic Energy of Spacecraft @8Km/Sec
20 Tonne Payload	0.13 Meter (13 Cm) Diameter Superconducting Cable	1.5 Meter OD Spacecraft Structure
8 Tonne Spacecraft Structure Weight	HTS Superconductor @ 4.2 Degrees Kelvin.	1.8 Meter OD Sabot around Star Tram Spacecraft
28 Tonne Total Launch Weight	12 Tesla Maximum Magnetic Field @ SC Cable Surface	11.4×10^6 Newtons Propulsion Force @40g.

8 km/sec Launch Velocity	35,000 amp/cm2 Operating Current Density in SC Cable	4 Million Amp Turn Current in Superconducting Cables on Spacecraft
40 g (400 m/sec^2) Acceleration Rate	10 Centimeter Gap Between Spacecraft and Tube Wall	

FIGURE 3.7.

LAYOUT OF THRUSTER CABLE AND SLAB RELATIVE TO SUPERCONDUCTING CABLE ON STARTRAM SPACECRAFT

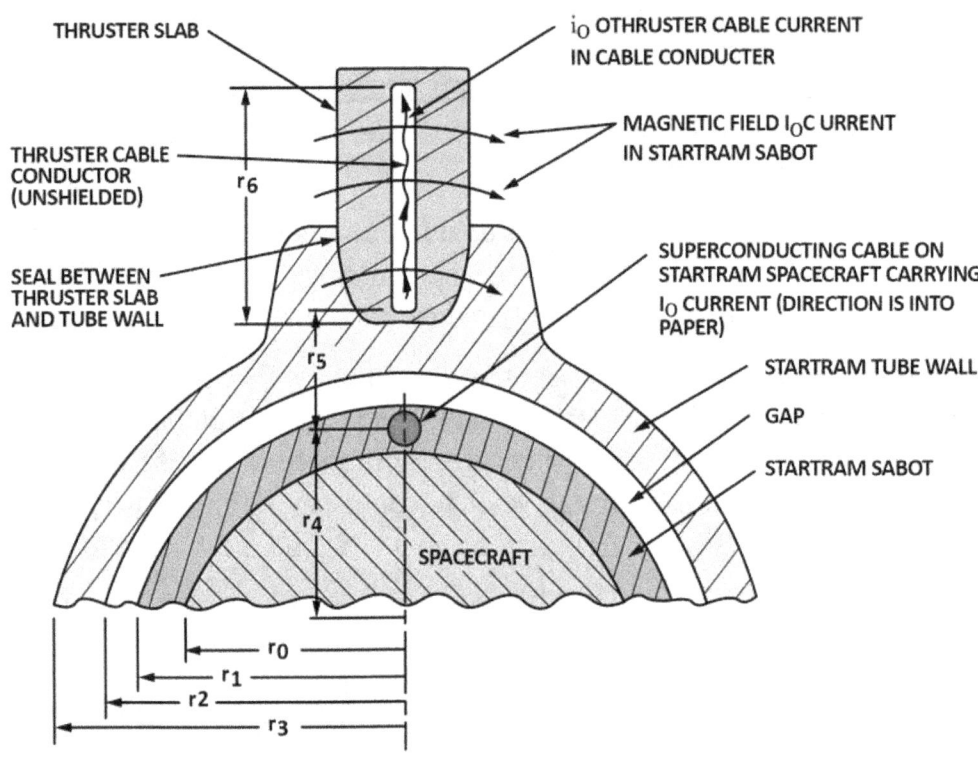

r_0 = radius of StarTram spacecraft

r_1 = radius of StarTram sabot

r_2 = inner radius of vacuum tube wall

r_3 = outer radius of vacuum tube wall

r_4 = radial distance of center of superconducting cable carrying IO current

r_5 = radial distance from center of superconducting cable to base of thruster cable

r_6 = radial distance from base of thruster cable to top of thruster cable

Note: For the directions of the Io and io currents shown in the diagram, the magnetic propulsion force on the spacecraft is out of paper

The 4 thruster circuits are encased in epoxy-fiberglass (or non-conductive material) slabs that can be inserted into slots in the wall of the evacuated launch tube. They do not pierce the wall of the launch tube, so that air cannot leak into the evacuated tube. They are tightly sealed to the tube wall and supported by it, to transmit the forces on the thruster cables, which are directed along the thruster slab. The thrust force on each of the 4 slabs in 1/4th of the total propulsion force of 11.4×10^6 Newtons on the StarTram spacecraft and directed in the opposite direction from the forward magnetic thrust on the spacecraft.

As the Spacecraft proceeds along the launch tube, the forces on the 2 thruster cables that act on it also travel along the thruster slab.

Table 3.2 gives the parameters for the thruster cable circuits imbedded in the 4 thruster slabs operating in each launch tube Section. The launch tube Sections are 100 meters in length, with 800 Sections forming the 80-kilometer-long launch tube, as shown in Figure 3.8.

TABLE 3.2.

DESIGN PARAMETERS FOR THRUSTER CABLE CIRCUITS IN ILLUSTRATIVE GEN-1 STARTRAM DESIGN

100-Meter-Long Launch Tube Section	25 MJ Inductive Energy in 100 Meter Thruster Circuit
4 Thruster Circuits, Each 100 Meters in Length, Positioned At 90 Degree Intervals Around The Launch Tube	0.5 Seconds to Reach Full Thruster Current of 1 Million Amps Using MACE Energy Storage Unit
Thruster Conductor Is Copper Cable At 300 K, With 200 cm² Area	275 MJ Propulsion Energy Delivered to Spacecraft Per 100 Meter Thruster Circuit
40 Thruster Cables, Each 1 Meter Long Exposed To Magnetic Fields From DC Superconducting Cables On StarTram Spacecraft	0.1 Second Dwell Time at Full Thruster Current Before Spacecraft Enters 100 Meter Long Thruster Section at 0.6 Second after start of MACE current discharge
Remaining 240 Meters of Thruster Cable Enclosed In Hollow Conducting Tubes To Shield Them From Transient Magnetic Field From Spacecraft	7.5 Degrees Kelvin Temperature Rise in Thruster Copper Conductor at 0.6 Second Point
280 Meter Total Thruster Cable Length Per Thruster Circuit	Residual Current in Thruster Circuit Dumped into External Shunt Resister After Spacecraft Passes
Thruster Cables Spaced 2.5 Meters Apart in each 100-Meter Section	17 Degrees Kelvin Total Temperature Rise in Resistor Conductor After MACE Current Goes to Zero
2 Thruster Cables Act On Spacecraft per Thruster Circuit	240 MW Peak I^2R Power In Thruster Circuit

8 Thruster Cables Act On Spacecraft for 4 Thruster Circuits	1 Million Amps Current in Thruster Cable Circuit
22,000 Megawatt Peak Propulsion Power Per Thruster Circuit@8km/sec	5×10^{-5} Henry Inductance of Thruster Circuit (40 Meter Shielded Cable; 40 Meter Exposed Cable)
1.73×10^{-6} Ohm Cm In Resistivity of Copper Conductor	
2.4×10^{-4} Ohm Electrical Resistance of 280-Meter-Long Cable Per Thruster Circuit	

Each 100-meter section has its own SMES energy storage unit that supplies the pulsed DC thruster currents that magnetically propel the StarTram spacecraft. The particular design for the SMES unit, termed MACE (Magnetic Cable Energy) is described in more detail later.

Summarizing, MACE consists of a circular loop of thermally insulated primary winding of High Temperature Superconductor (HTS) enclosed by an outer winding of ordinary copper conductor. As the spacecraft approaches a given 100-meter launch tube section, the current in the primary HTS superconductor winding is inductively transferred to a secondary winding in a fraction of a second, e.g., 1/2 second.

The current that had been flowing in the primary HTS winding then begins to flow in the secondary copper winding by transformer action. The coupling efficiency is very high—more than 95% of the magnetic energy that was stored in the MACE loop when its current flowed in the primary HTS superconductor winding is transferred to the secondary copper winding.

The secondary copper winding in the MACE energy storage unit is connected to the thruster cables in the 100-meter long thruster slab, providing the pulsed DC currents that flow in the thruster cables.

As illustrated in Figure 3.8, there are 800 launch tube sections each 100 meters in length, along the 80-kilometer length of the evacuated launch tube. The acceleration time to reach 8 km/sec is 20 seconds at 40 g.

The transit time to pass through a 100-meter launch tube section depends on spacecraft velocity, being 0.1 second at a velocity of 1 km/sec, 0.025 second at 4 km/sec, and 0.0125 second at the final 8 km/sec.

The propulsion force of 11.4×10^6 Newtons, is constant with velocity, but the propulsion power increases linearly with velocity being 11 gigawatts at 1 km/sec, 44 gigawatts at 4 km/sec, and 88 gigawatts at 8 km/sec.

Each 100-meter section adds 1.1 gigajoules of kinetic energy to the spacecraft. As velocity increases, however, the ΔV increase per 1.1 gigajoules addition decreases.

In the first 1-kilometer-long section of the launch tube, velocity increases from 0 Km/sec to 0.9 Km/Sec. In the 1-kilometer-long section at 39 Km, velocity increases from 5.58 Km/sec to 5.65 Km/sec. In the last 1-kilometer-long section at 79 Km, velocity increases from 7.95 Km/sec to 8 Km/sec.

Of the 280-meter-long thruster cable in a 100-meter-long section, only 40 meters is directly exposed to the magnetic field from the superconducting DC cables on the space craft. As illustrated in Figure 3.6, 240 meters is inside hollow conducting tubes that shields them and the current they carry from the spacecraft's magnetic field.

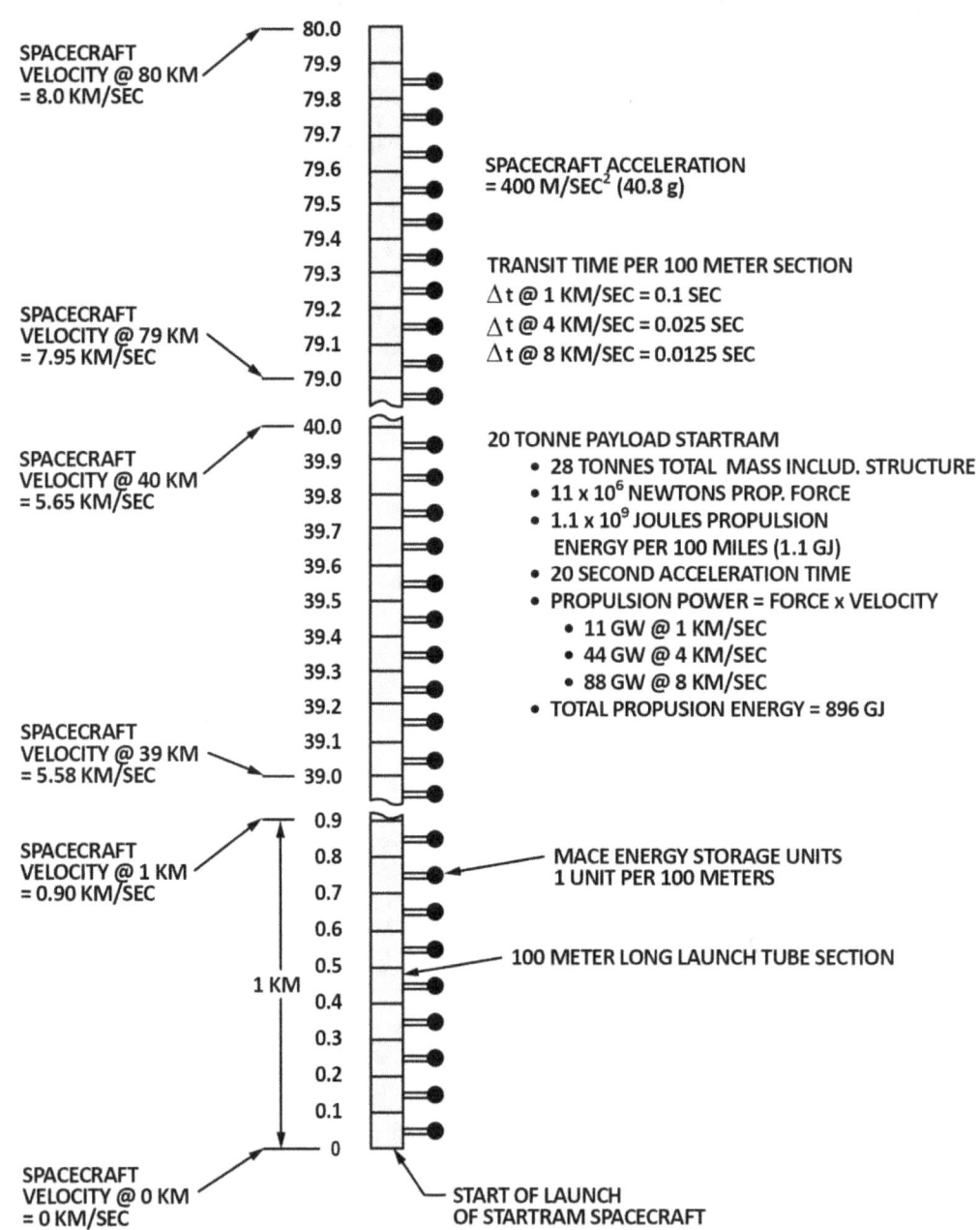

FIGURE 3.8.

LAYOUT OF STARTRAM EVACUATED LAUNCH TUBE FOR SPACECRAFT ACCELERATION TO LAUNCH VELOCITY

The hollow tubes are made of electrically conducting material, e.g., copper or aluminum, with a sufficiently thick wall that the pulsed external magnetic field as the spacecraft passes does not penetrate the tube. Similar shielding of cables is common in the power and communications industry.

As illustrated in Figure 3.6, exposed thruster cables descend radially inwards toward the spacecraft (bottom of the thruster slab in Figure 3.6). After reaching the bottom of the thruster slab, they then return to the top of the slab and the thruster circuit inside a shield tube.

The shielding prevents the cable current inside the shielded tube from magnetically interacting with the magnetic field from the spacecraft, since if it did interact, the magnetic force on it would oppose the magnetic force from the exposed thruster cable, resulting in zero net propulsion force.

The time internal that the shielded cables is exposed to the magnetic field from the spacecraft is extremely short, only a few milliseconds. For a 5-meter-long magnetic field on the spacecraft traveling at 5 Km/sec, the shield tube experiences the spacecraft's magnetic field for only 1/1000th of a second. Exposure time is even shorter at higher speeds, and a bit longer at lower speeds.

The transient exposure of the shield tube to the spacecraft's magnetic field produces eddy currents in the shield that prevent the external field from penetrating the shield. The transient eddy currents produce a transient magnetic force on the shield tube and a corresponding magnetic force on the spacecraft, but the net propulsion force on the spacecraft during the spacecraft passage is zero, except for a very small magnetic drag, due to I^2R energy losses from the eddy currents in the shield.

The total length of the tubes that shield the thruster cables in the 100-meter-long thruster cable circuit is only 40 meters, the same length as that for the exposed thruster cables.

There is another 100 meters of shielded tube at the top of each thruster slab inside of which the go and return thruster cables are located (Figure 3.6). Every 2.5 meters, a thruster cable exits from the shield tube and travels radially inwards inside a shield tube for 1 meter towards the center of the launch tube.

At the bottom of its 1-meter length, the thruster cable turns and travels radially upwards and enters the 100-meter-long shield tube at the top of the thruster slab. (Figure 3.6)

The magnetic field strength from the StarTram's superconducting DC cable at the top of the thruster slab is

$B_{TOP} = \left(\frac{\mu o}{2\pi}\right) I_o$ amp/1.2 meter = 0.67 Tesla

A given point on the 100-meter-long shield tube at the top of the thruster slab will experience a traveling external magnetic field pulse of 0.67 Tesla magnetic field strength for a time interval of

Δt_{top} = 5 meters/$V_{spacecraft}$

Where 5 meters is the length of the SC cable on the StarTram spacecraft and $V_{spacecraft}$ is its velocity. At $V_{spacecraft}$ = 5 km/sec

Δt_{top} = 5/5000 = 1/1000th of a second

The magnetic force and I^2R heating effects on the shield tube from eddy currents will be very small.

FIGURE 3.9.
DRAWING OF MACE CIRCULAR LOOP ENERGY STORAGE UNIT

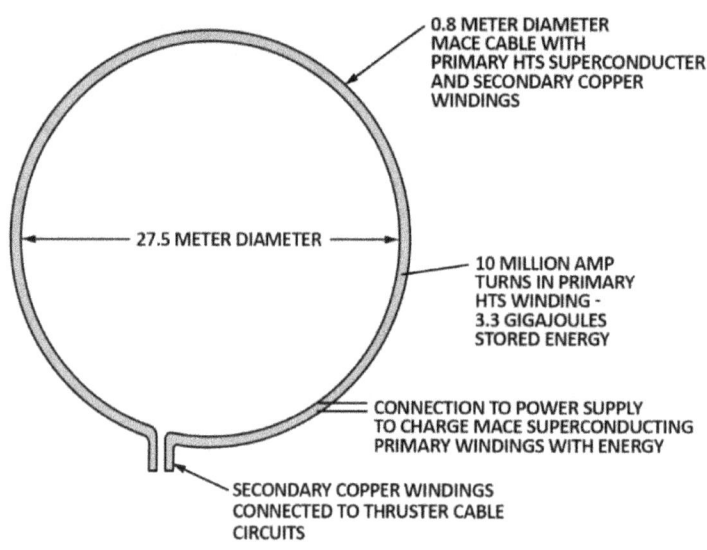

The 4 thruster cable circuits in the 100-meter-long thruster slab obtain their 1-million-amp current from a MACE (MAgnetic Energy Cable) energy storage unit located next to the 100-meter-long evacuated launch tube Section.

FIGURE 3.10.

SUPERCONDUCTING PRIMARY AND SECONDARY COPPER WINDING ON MACE CABLE

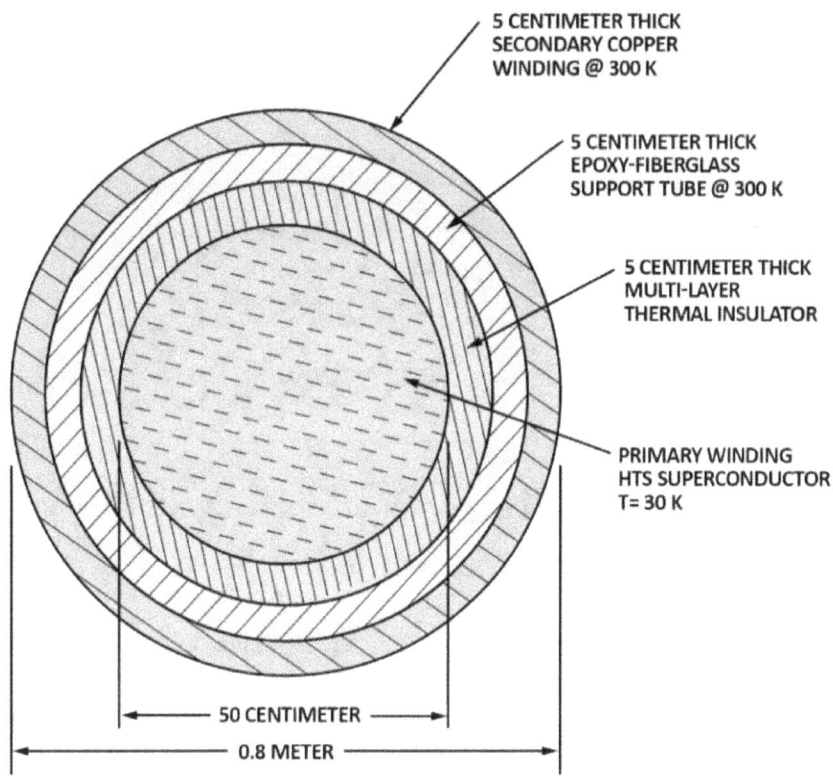

The MACE energy storage loop is a simple circular loop of thermally insulated superconductor cable (Figure 3.9) that carries very large current, on the order of 10 million amp turns in its multiturn winding. With a diameter of 27.5 meters, the energized MACE unit stores 3.3 gigajoules of magnetic energy, 3 times greater than the 1.1 gigajoules of propulsion energy imparted to the StarTram spacecraft as it passes through the 100-meter-long launch tube section.

The MACE cable has 2 windings, as illustrated in Figure 3.10. The inner primary winding is a multiturn High Temperature Superconducting coil that carries a total current of 10 million amp turns (MAT). The HTS superconductor is commercial YBCO superconductor (Figure 3.11 and 3.12). It operates at 30 degrees Kelvin using commercially manufactured cryocoolers for refrigeration.

The primary HTS winding is thermally insulated with a layer of multi-layer vacuum insulation to minimize thermal leakage into the primary winding. Outside of the thermal insulation layer there is a composite tube wall on which is wound a secondary winding of copper conductor at ambient temperature of 300 degrees Kelvin (27 degrees Celsius)

FIGURE 3.11.
DRAWING OF SUPERPOWER YBCO SUPERCONDUCTING TAPE

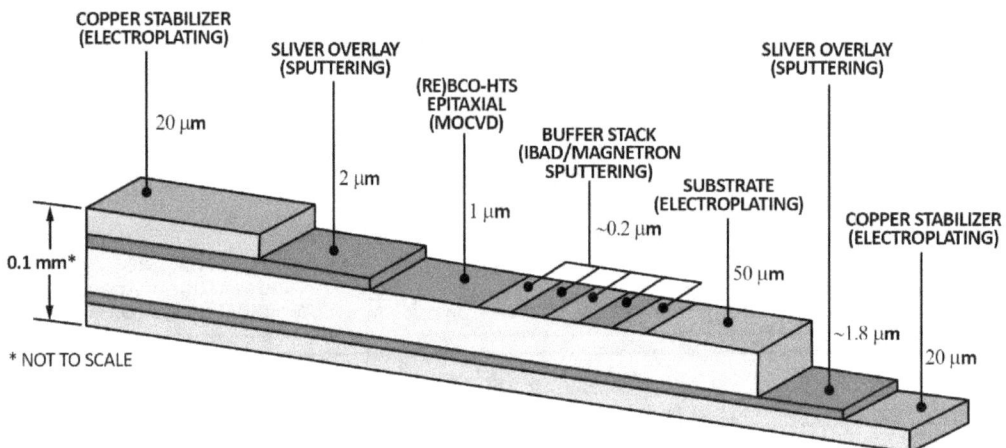

SuperPower's (RE) BCO superconductor with artificial pinning

Structure provides a solution for demanding applications:

Hastelloy® C276 substrate

high strength

high resistance

non-magnetic

Diffusion barrier to metal substrate

Ideal lattice matching from substrate through REBCO

MOCVD grown (RE)BCO layer with BZO nanorods

Flux pinning sites for high in-field Ic

Silver and copper stabilization

FIGURE 3.12.
GRAPHS OF CRITICAL CURRENT OF SUPERPOWER YBCO SUPERCONDUCTING TAPE AS FUNCTION OF MAGNETIC FIELD STRENGTH AND TEMPERATURE

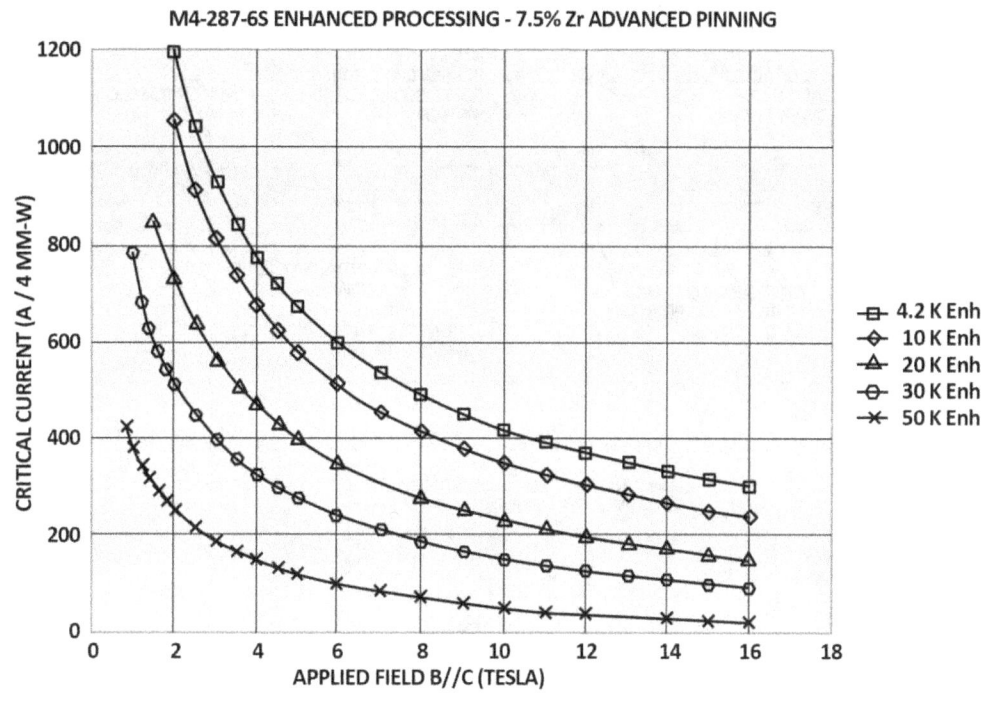

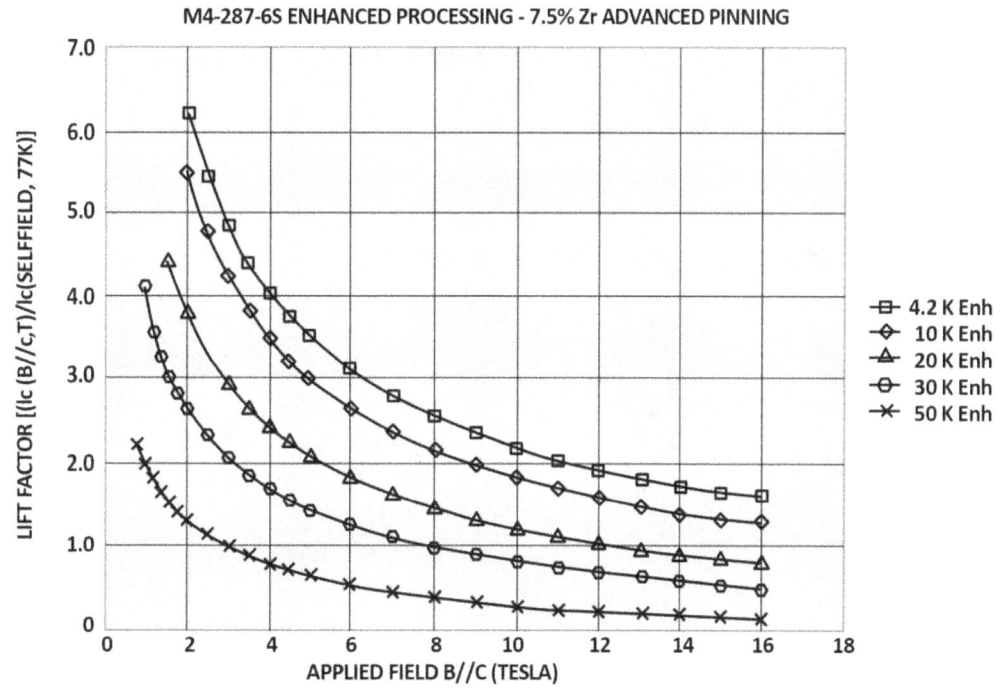

MEASUREMENTS MADE AT TOHOKU UNIVERSITY

FIGURE 3.13.
GEOMETRY AND OPERATION OF MACE ENERGY STORAGE UNIT WITH THRUSTER CABLE CIRCUITS

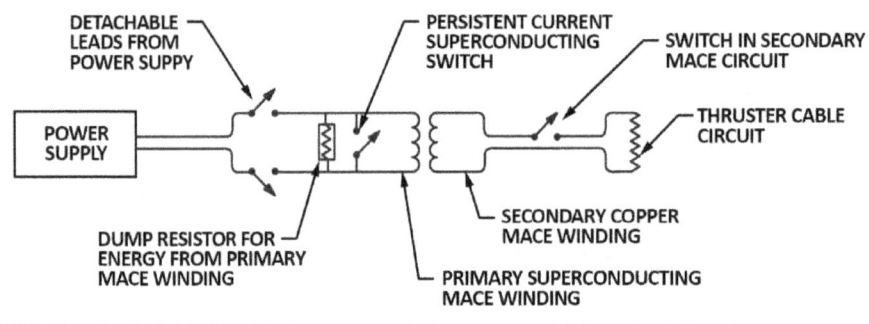

STEP 1: CHARGE MACE ENERGY STORAGE UNIT

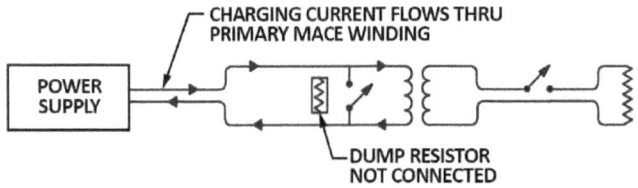

STEP 2: MACE UNIT FULLY CHARGED AND IN PERSISTENT CURRENT MODE

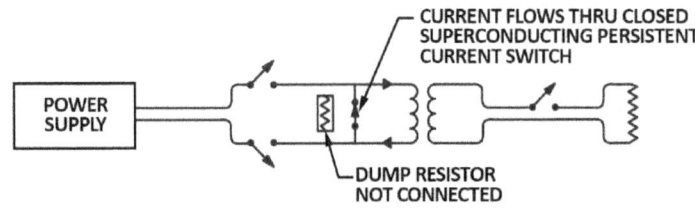

STEP 3: MACE UNIT BEGINS DISCHARGE TO THRUSTER CABLE CIRCUIT

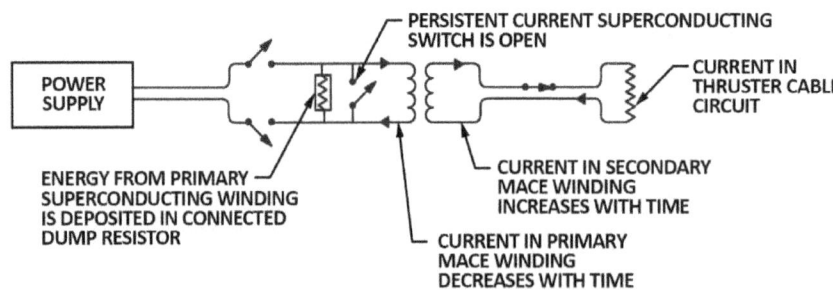

Figure 3.13 illustrates how the MACE unit is charged with magnetic energy, and how it discharges its magnetic energy into the thruster cable circuits when it accelerates the spacecraft passing through a 100-meter-long launch tube section.

The MACE unit has the following components:

DC power supply that energizes the multi-turn HTS primary winding

The MACE cable (Figure 3.10) with a primary superconducting HTS winding at 30K and a enclosing ordinary copper winding at 300 K, and thermal insulation layer between them

Dump resistor that receives the magnetic energy from the primary HTS winding, reducing its current to zero

Connection of the secondary copper winding to the 4 thruster cable circuits in the 100-meter-long evacuated launch tube

Switches that control power flow from the MACE winding, delivering it where and when it is required

There are 3 operational steps for the MACE unit, as illustrated in Figure 3.13.

Step 1: Charging the MACE Storage Unit

The HTS Superconducting winding is at 30 degrees Kelvin and zero current. No magnetic energy is stored. Demountable leads from DC power supply are attached to the ends of the superconducting HTS primary winding by closing switches, and the current in the winding increased from zero to its full operating current. The other switches in the MACE System – the dump resistor switch, the persistent current switch and the switch in the secondary copper winding are open.

Step 2: When the current from the power supply reaches the value that corresponds to a total of 10 million amps turns in the multi-turn primary HTS superconducting winding, the persistent current superconducting switch is closed and the switches to the power supply opened. The number of turns used in the primary HTS winding depends on the current level of the power supply. For example, with 10,000 amps from the power supply, and 10 million amps turns in the HTS primary winding, the HTS winding would have 1000 turns of an HTS tape bundle. No current in the secondary copper winding because its switch was open during the charging process.

The charging process can be very slow since there is no I^2R losses in the superconducting primary HTS winding, charging a MACE unit with 3.3 Gigajoules of magnetic energy over a 10-hour period takes only a 100-kilowatt power supply. Charging the full launch tube system with 800 MACE units, over a 10-hour period would take 80 megawatts, a modest amount compared to conventional power plants that generate 1,000 megawatts.

Step 3: Discharging the stored magnetic MACE energy to the thruster cable circuits. The secondary copper winding on the MACE loop carries no current in Step 2. When the primary HTS current is decreased, current flows in the secondary copper winding by inductive action. In effect the MACE circular loop is a transformer. As current drops in the primary winding, inductive coupling causes current to increase in the secondary winding.

The coupling between the MACE primary and secondary coils is very close. When the primary HTS current goes to zero, approximately 95% of the stored magnetic energy transfers to the secondary copper winding, with its total amp turns only slightly less than the 10 million amp turns stored in the primary winding during step 2.

MACE-like energy storage units are not new. There has been extensive R&D on pulsed superconducting transformers, using an approach in which the superconducting primary windings is suddenly driven above its transition temperature by pulsed heaters inside the winding. The current that had flowed in the winding then transfers to a secondary copper winding.

This approach, while feasible, has the disadvantage that the primary superconducting winding temperature increases considerably, and must be re-cooled using extensive refrigeration equipment and power over a long period before the storage unit can operate again.

MACE avoids this problem by dumping the current from the primary winding into an external dump resistor (Figure 3.13) that operates at room temperature. The MACE primary winding stays at 30K and does not have to be re-cooled.

The secondary copper winding is wound using the 4 separate circuits of 2 turns per circuit around the MACE circular loop. Each of the 4 separate circuits is connected to one of the for-thruster lab circuits in the 100-meter-long evacuated launch tube Section.

Approximately 9 Mega Amp Turns (MAT) of total current flows in the secondary winding after 90% of the current originally in the HTS primary winding has been discharged into the dump resistor. With 8 turns in the secondary winding, each of the 4 thruster cable circuits has a current of 9/4x2=1.1 million amps.

Figure 3.14A shows the total currents in the MACE primary and secondary windings as a function of time for a dump period of 0.5 seconds, in which the primary current has decreased from its full current of 10 million amps turns down to 1 million ampturns by I^2R losses in a dump resistor. During the 0.5 second dump period the secondary winding current increases from zero to 9 million amps turns and the current in each of the 4 thruster cable circuits increases to 1.1 million amps (Figure 3.14B).

FIGURE 3.14A.
TOTAL CURENT IN MACE PRIMARY & SECONDARY WINDINGS AS FUNCTION OF TIME FOR 100-METER-LONG SECTION OF LAUNCH TUBE

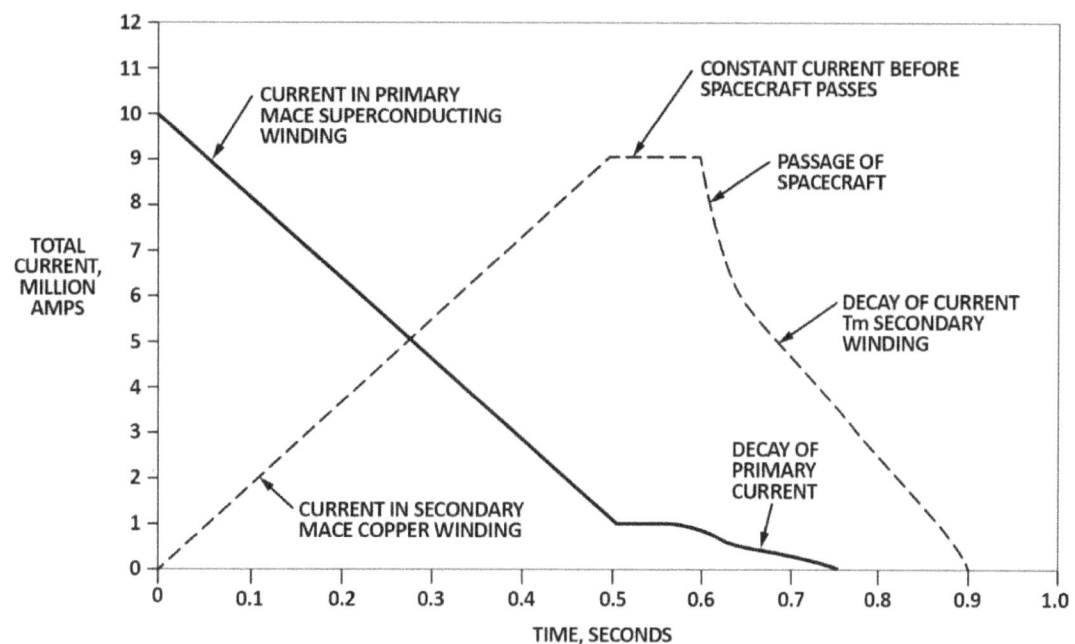

FIGURE 3.14B.
CURRENT IN THRUSTER CABLE CIRCUIT AS FUNCTION OF TIME FOR 100-METER-LONG SECTION OF LAUNCH TUBE

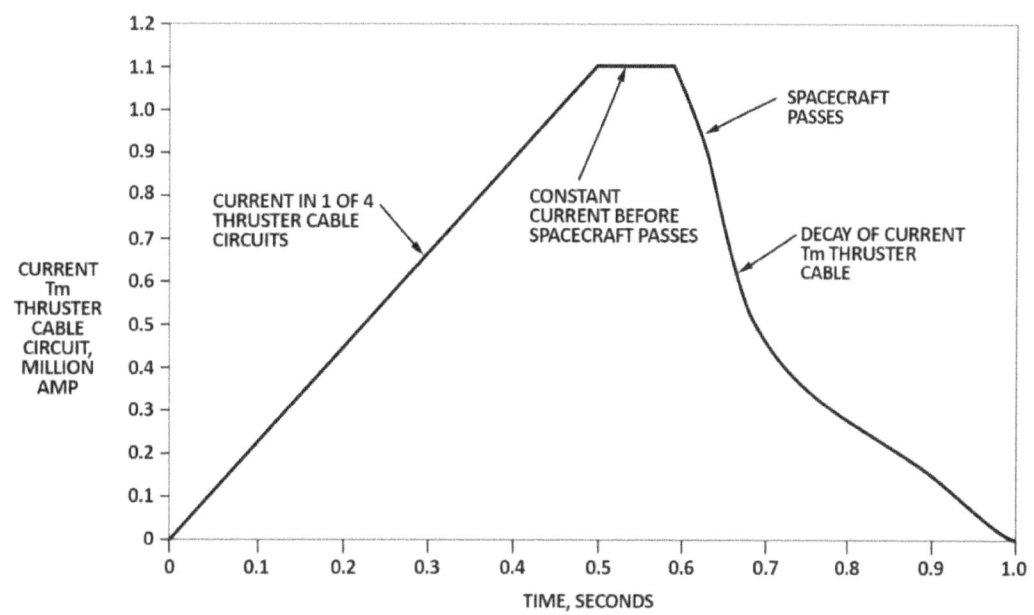

FIGURE 3.15A.

VOLTAGE AND CURRENT IN 1 THRUSTER CIRCUIT AS A FUNCTION OF TIME TILL SPACECRAFT PASSES

Basis:
1 Thruster circuit out of 4 total
100-meter-long launch tube section

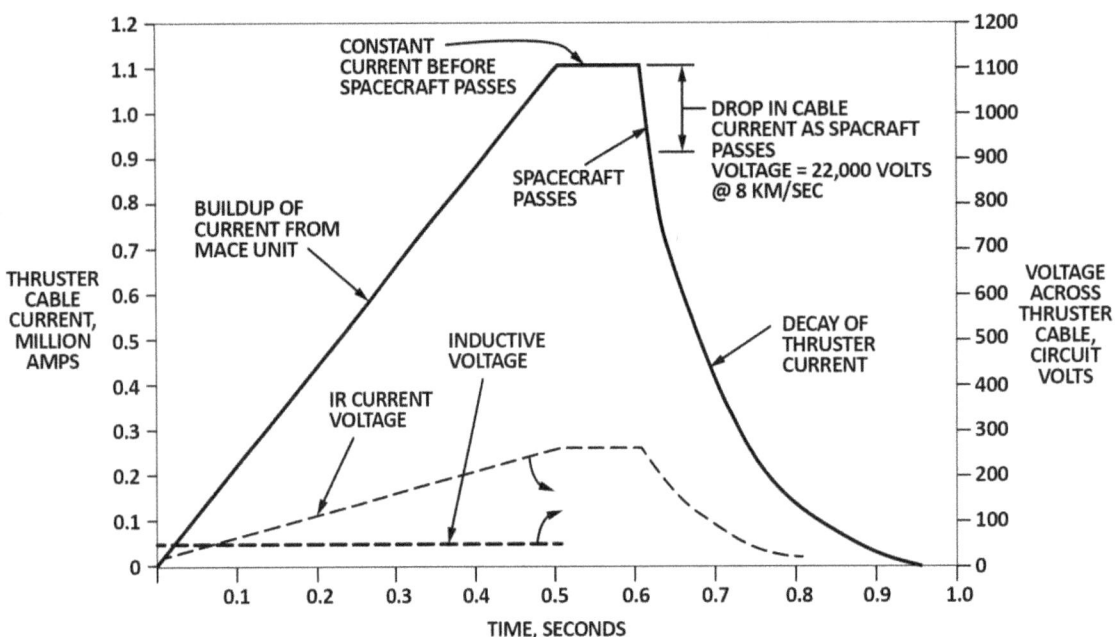

After reaching full current there is a short interval, e.g., 0.1 second, of virtually constant current until the spacecraft reaches the 100-meter-long energized launch tube Section. The current in the secondary MACE winding and the thruster cables begins to decrease as the spacecraft moves along the 100-meter Section, with stored magnetic energy in the thruster cable circuits and secondary winding being transferred as propulsion energy to the accelerating spacecraft.

For the MACE/StarTram design described here, the initial stored energy in the thruster circuits and secondary winding as the spacecraft enters the 100-meter Section is 3.3 Gigajoules,

As the spacecraft leaves the 100-meter section, 1.1 Gigajoules has been transferred to the spacecraft as propulsion energy, leaving 2.2 Gigajoules of energy in the MACE circuit, resulting in a current decrease from i_0 down to i_1,

The ratio of final current, i_1 to initial current, i_0 is given by:

$$\left(\frac{E_1}{E_0}\right)^{\frac{1}{2}} = \left(\frac{2.2}{3.3}\right)^{\frac{1}{2}} = 0.82$$

For $i_0 = 1.1 \times 10^6$ Amp, $i_1 = 0.90 \times 10^6$ Amp and the average current is approximately,

$I_{avg} = \frac{i_0 + i_1}{2} = 1 \times 10^6$ Amp

The design value for the thruster circuits.

Figures 3.15A and 3.15B show the voltage, current, and energy in the thruster cable circuit during the buildup of current in the 0.5 sec internal discharge of the primary HTS superconductor current, the 0.1 second interval before the spacecraft enters the 100-meter launch tube Section, the very short interval as it passes through the 100-meter Section, and after it leaves the section.

Each of the 4 thruster circuits transfers 275 megajoules to the spacecraft as it passes through the 100-meter Section, for a total transfer of 1.1 gigajoules. The IR and inductive voltages during the buildup of the thruster current are small, a few hundred volts at peak, and the corresponding energy requirements, i.e., I^2R losses and inductive energy are very modest compared to the propulsion energy.

As the spacecraft passes through the 100-meter Section, however, the voltage across the thruster cable circuit increases greatly, due to the high velocity of the spacecraft and its strong magnetic field.

FIGURE 3.15B.
ENERGY IN 1 THRUSTER CIRCUIT AS A FUNCTION OF TIME TILL SPACECRAFT PASSES

Basis:
1 Thruster circuit out of 4 total
100-meter-long launch tube section

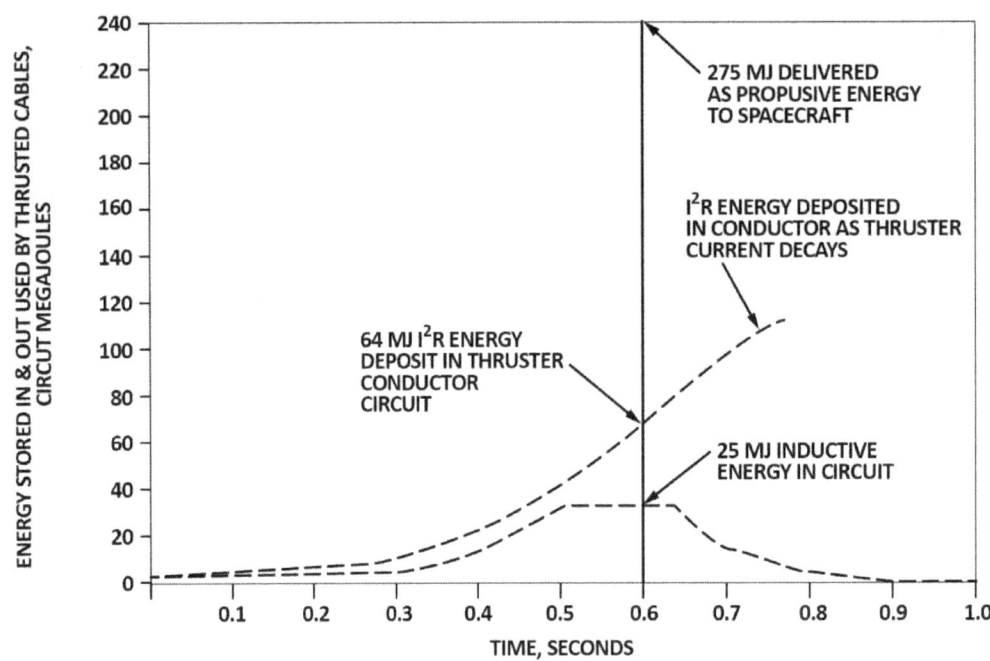

The product of thruster voltage times the thruster cable current equals the propulsion power applied to the spacecraft. At maximum velocity of VS = 8 Km per second and a propulsion force of 11×10^6 Newtons, propulsion power from the 4 thruster cable circuits is:

Pp = Fp x Vs = 11×10^6 x 8,000 = 88 Gigawatts

With the power level in each of the 4 thruster cable circuits equal to 88/4 or 22 Gigawatts. The voltage V_c across the circuit is

V_c = (Pp/4) i_o = 22,000 volts

Where i_o = thruster current in the circuit = 1×10^6 Amp. The voltage appears across the 2-exposed thruster cables that are interacting with the magnetic field from the spacecraft, with 11,000 volts across each of the cables. As the spacecraft moves along the 100-meter launch tube Section, the locations of the thruster cables that experience 11,000 volts move along with it.

At 8 km/sec, 2.5-meter spacing between exposed thruster cables, and a 5-meter-long magnetic field region or the spacecraft, a given exposed thruster cable experiences the 11,000 volts across it for only 0.000625 second.

After the spacecraft leaves the 100-meter long launch tube Section, it has acquired 1.1 gigajoules of additional propulsion energy, 1/3rd of the 3.3 gigajoules originally stored in the MACE unit that provides energy for the 100-meter Section 2/3 of the original MACE energy, i.e., 2.2 gigajoules remain in the MACE secondary copper winding and the 4 thruster cable circuits, decaying at the rate determined by the inductance and resistance of the secondary winding and thruster cables.

Figures 3.16A and 3.16B show 2 options for managing the current/energy decay and the resulting rise in temperatures of the copper conductors in the circuit.

In option 1, the remaining current continues to store in the secondary winding and thruster cable conductors, reaching zero after several seconds. For the illustrative StarTram design presented here, the resulting temperature in the copper conductors would be approximately 357 degrees Kelvin (84 degrees Centigrade).

This appears acceptable, with the conductors reaching a maximum temperature of 357K. The conductors would then be cooled down to 300K (27 °C) over the next few hours to be ready for the next launch.

One may ask, why is only 1/3 of the MACE energy extracted for propulsion energy? Why not 2/3 or even more? This would reduce the amount of residual energy in the propulsion circuits, and decrease their rise in temperature after I^2R decay of the current.

The answer is that extracting 1/3 of the stored energy enables the thruster current to be relatively constant during the time that the thruster cables accelerate the spacecraft. Their initial current of 1.1 million amps when the spacecraft enters the 100-meter launch tube section drops to 0.9 million amps as it leaves the Section for an average current of about 1 million amps. Extracting a greater fraction of the MACE energy would result in a much greater variation in thruster cable current during the acceleration interval.

Option 2 (Figures 3.16A and 3.16B) provides a way to reduce the temperature rise in the copper conductors of the secondary winding/thruster cables. In option 2 after the spacecraft leaves the 100-meter launch tube Section, a shunt resistor is connected into the overall circuit in parallel with the thruster cables. Its electrical resistance is much smaller than the electrical resistance of the thruster cables so that the cable currents and I^2R losses are greatly diminished after the spacecraft passes.

As a result, the temperature rise, for the example shown, in the thruster cables is reduced from 57K to 17K (Figure 3.16B) and their final temperature is 317K(44°C). The total amount of energy dumped is the same as in option 1, i.e., 2.2 gigajoules, but most of it now goes into the shunt resistor and not the thruster cables.

FIGURE 3.16A.

OPTIONS FOR DECAY OF CURRENT IN THRUSTER CABLE CIRCUIT AFTER SPACE CRAFT PASSES

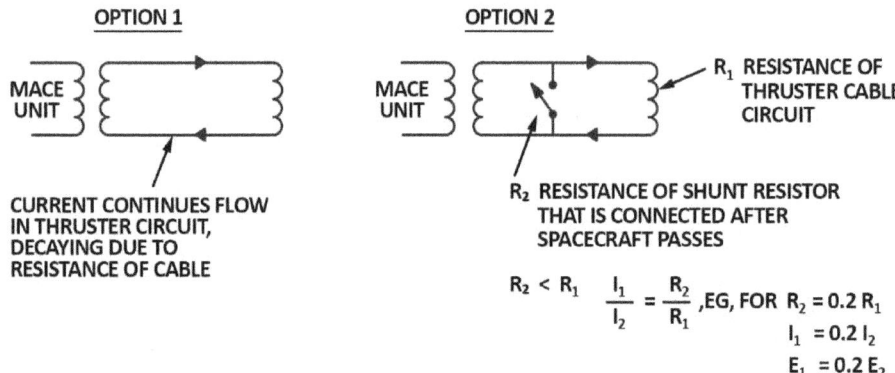

FIGURE 3.16B.

TEMPERATURE RISE IN THRUSTER CABLE CIRCUIT FOR OPTIONS 1 AND 2

Basis:

3.3 GJ total energy in unknown unit (4 thruster circuits)

825 MJ in each thruster circuit

275 MJ propulsion energy delivered to spacecraft

550 MJ dissipated as I^2R energy in conductors

60 MJ dissipated as I^2R energy in thruster circuit before spacecraft passes

Option 1:

490 MJ dissipated in thruster circuit after spacecraft passes

Option 2:

82 MJ dissipated in thruster circuit after spacecraft passes

408 MJ dissipated in shunt resister after spacecraft passes

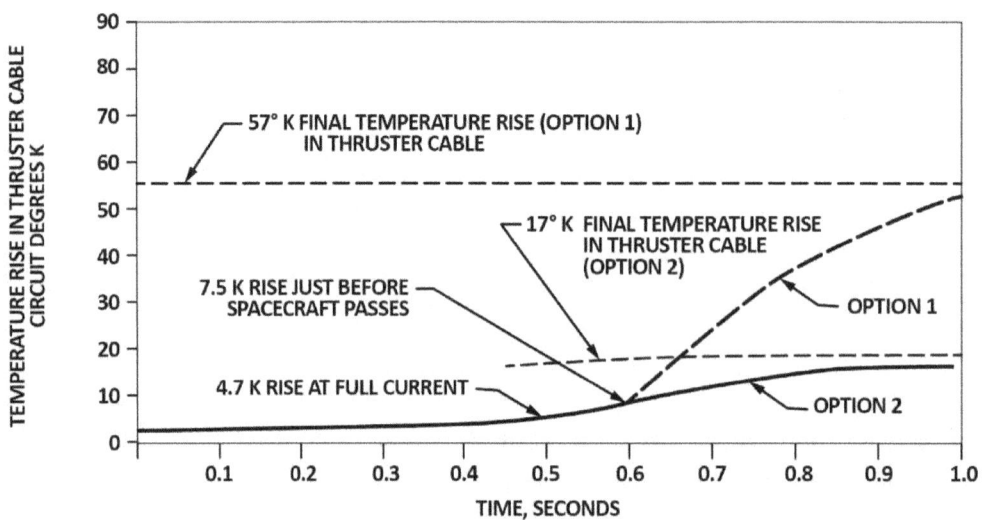

The MACE energy storage system is described in detail in a separate report. Table 3.3 summarizes its principal features.

TABLE 3.3.
DESIGN PARAMETERS FOR MACE ENERGY STORAGE UNIT

1 MACE unit per 100-meter-long launch tube Section	0.0125 second time for spacecraft to pass through 100-meter Section at 8 Km/sec.
MACE unit supplies 1-million-amp average current to each of 4 thruster cable circuits in 100-meter launch tube Section	2/3 of original MACE stored energy appears as I^2R losses (2.2 GJ) in copper windings and shunt resistor after spacecraft leaves 100-meter Section.
1 MACE unit stores 3.3 gigajoule	Over 95% of MACE initial magnetic energy transfers to secondary copper winding circuit when primary HTS current is dumped.
1 MACE unit supplies 1.1 gigajoules propulsion energy to spacecraft	MACE primary winding current is 10 million amp turns
MACE has primary winding of High Temperature Superconductor (HTS) at 30K, secondary inductively coupled copper winding at 300K.	Initial current in each thruster cable circuit is 1.1 million amps, before spacecraft passes, decreasing to 0.9 million amps after passage.
MACE activated when current in primary superconducting winding is shunted to dump resistors. MACE current and energy then inductively transfer to secondary copper winding, which is connected to thruster cable circuits.	0.5 seconds to reach full thruster cable current from start of MACE activation followed by 0.1 second dwell time before spacecraft enters 100-meter Section.
MACE geometry is circular loop 27.5 meters in diameter	

The StarTram spacecraft is automatically vertically and laterally stabilized as it moves along the evacuated launch tube by magnetic interaction between stability loops of ordinary conductor, e.g., copper or aluminum, that are located in the thruster slabs and the HTS superconducting cables on the levitated spacecraft.

Figure 3.17 shows the geometric arrangement of the vertical and stability loops and the superconducting cables on the spacecraft. The vertical stability loops A and B are connected in a null flux circuit. When the spacecraft is centered in the launch tube, the magnetic flux through vertical loop A is equal and opposite to the magnetic flux through vertical loop B, with the result that the net magnetic flux through the AB circuit is zero, and no current flows in the circuit (Figure 3.18A).

If the spacecraft is displaced vertically towards loop A, it has a higher value of magnetic flux than loop B because it is closer to the spacecraft superconducting cable.

As a result, the net magnetic flux through the AB circuit is non-zero, and a current is induced in the circuit. The current flows in a direction so that the magnetic flux from it equals and opposes the net magnetic flux from the current carried in the superconducting cables on the spacecraft.

Figure 3.17

VERTICAL AND LATERAL STABILITY OF STARTRAM SPACECRAFT

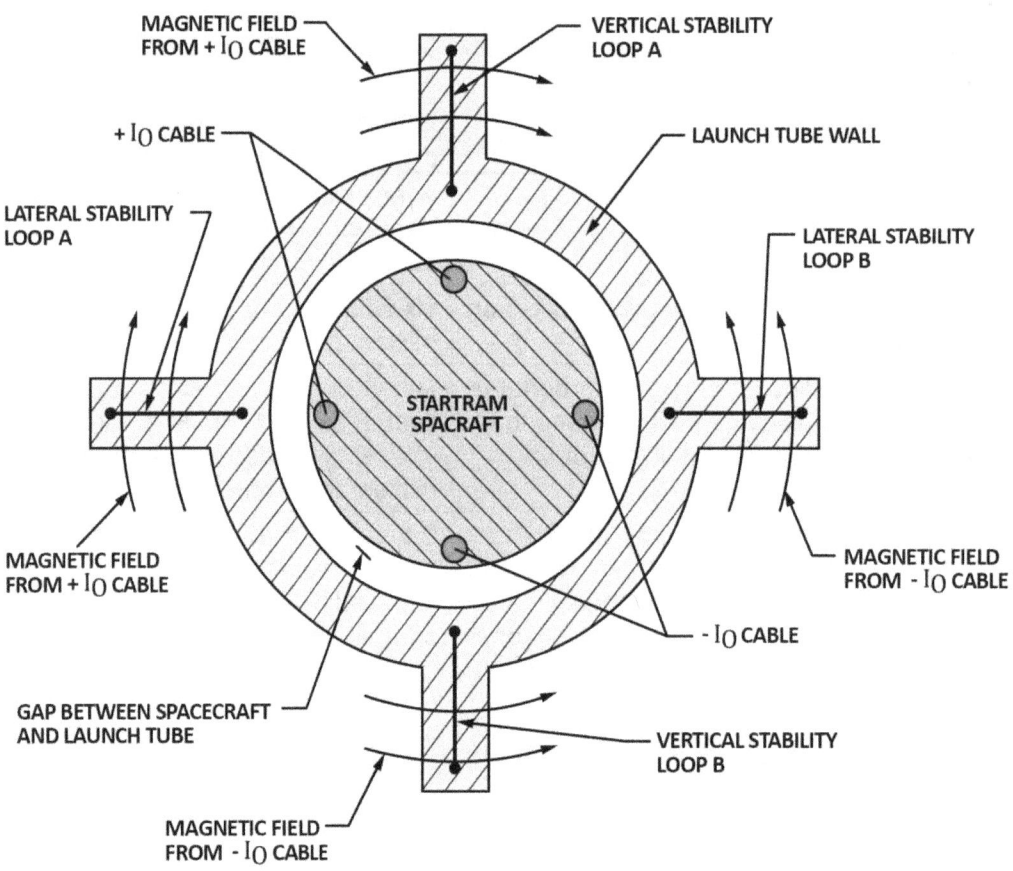

The current in vertical stability loops A and B magnetically interacts with the superconducting cables on the spacecraft (Figure 3.18B) pushing it back towards the centered position.

The same type of reaction occurs if the spacecraft is displaced towards vertical stability loop B (Figure 3.18C). There is a net magnetic flux through the AB current and an induced current is. In this case, however, the magnetic force between it and the superconducting cables on the spacecraft is in the opposite direction, pushing it towards loop A to center it, rather than away from loop A.

The process is similar for the lateral stability loops. The induced currents flow in a direction that pushes the spacecraft back towards the center of the launch tube if it is displaced from the center of the tube.

The stability loops are located between the thruster cables on the thruster slabs which are spaced 2.5 meters apart along the thruster cable. As an example, consider a 1.5-meter-wide stability loop positioned between adjacent thruster cables, with a loop length of 1 meter, the same length as the thruster cables, and a distance of 0.2 meters of the base of the stability loop from the center of the superconducting cable on the StarTram spacecraft, the same distance that the base of the thruster cables is from the superconducting cable.

When the spacecraft is centered the magnetic flux through vertical stability loops A&B is:

$\Phi A = \Phi B = 2 \times 10^{-7} \, I0 \, \ln\left[\frac{r1}{r0}\right]$ (w) = 2.15 Webers

Where $r_0 = 0.2$ meters, $r_1 = 1.2$ meters, w = 1.5 meters, and $I_0 = 4 \times 10^6$ Amp

When the spacecraft is centered $\Phi A - \Phi B = 0$ and no current flows in loops A & B.

When the spacecraft is displaced by the distance Δy towards loop A.

$\Phi A = 2 \times 10^{-7} \, I0 \, \ln\left[\frac{r1 - \Delta y}{r_0 - \Delta y}\right]$ (w)

$\Phi B = 2 \times 10^{-7} \, I0 \, \ln\left[\frac{r_0 + \Delta y}{r_0 + \Delta y}\right]$ (w)

For $\Delta y = 0.01$ meter, (1 centimeter)

$\Phi A = 2.2016$ Webers, $\Phi B = 2.1015$ Webers, and $\Phi A - \Phi B = 0.100$ Webers

The inductance of the loops is (10-centimeter diameter conductor)

$L_A = L_B = 2.4 \times 10^{-6}$ Henries (single turn)

With the circuit, inductance $L_A + L_B = 4.8 \times 10^{-6}$ Henries and the induced current i_0

= 20,900 amps

The restoring stability force F_S on the spacecraft from the current io in the AB loop circuit is:

$F_S = \frac{\mu 0}{2\pi} I0 \, i0 \, w \left[\frac{1}{r_0 + \Delta y} + \frac{1}{r_0 - \Delta y}\right] \cong 2\left(\frac{\mu o}{2\pi}\right)\frac{I0 \, i0 \, w}{r_0}$

$F_S = 4 \times 10^{-7} \times 4 \times 10^6 \times 20{,}900 \times 1.5/0.2 = 250{,}000$ Newtons

With two stability loop circuits on the thruster slab interacting with the superconducting cable on the spacecraft, the total restoring force is:

F_{TOT} = 500,000 Newtons per centimeter of displacement

For the 28 tonne StarTram spacecraft, this corresponds to a restoring g force of:

= 18 m/sec² (1.8g) per centimeter of displacement

For a 5-centimeter displacement (½ of 10 cm gap) the restoring force would be:

F_{TOT} = 2.28 x 10^6 Newtons per 5 centimeters of displacement with

= 81 m/sec^2 (8.1g) per 5 centimeters of displacement

The lateral stability restoring force would be the same.

A continuous vertical levitation force of 1g is necessary to keep the StarTram spacecraft levitated as it moves along the launch tube.

There are 2 options for this:

Option 1. The spacecraft travels slightly below the center of the launch tube at a distance Δy below the center for which the upwards vertical stability force on the spacecraft mass equals 1g. The Δy distance for this is small, e.g., (1/1.78) (1 centimeter) = 0.56 cm (1/5th of an inch). This appears very feasible.

Option 2. Install conductors in the thruster slabs, which would be energized at the same time as the thruster cables, that would magnetically interact with the superconducting cables on the spacecraft to levitate it when it was centered in the launch tube. The required current would be only about 10,000 amps, much smaller than the million-amp thruster cable current. Option 2 also appears very feasible. Like the thruster cable current, the levitation current would only flow for a second or so in the 100-meter launch tube Section.

The temperature rise in loops A and B will be very small, even if the restoring force is very strong. For a 5-centimeter displacement from the center of the launch tube, with an 8g restoring force and 89,000 amps of current flowing in the loop A and B conductors, the temperature rise is:

$$\Delta T = \frac{(89,000)^2 \text{ conductor resistivity X transit time}}{\text{conductor area X conductor heat capacity}}$$

With the copper conductor 10 cm in diameter, resistivity equal to 1.72 to 10^{-6} ohm centimeters, area equal to 78.5 cm^2, heat capacity = 1.52 J/cm^{3}K, delta t = 1 millisecond at a spacecraft velocity of 5 km/sec.

ΔT = 0.11 °K (0.11 °C)

A negligible temperature rise.

Depending on location of the StarTram launch facility, certain portions may involve high centripetal accelerations, e.g., a laterally or vertically curving tube, with accelerations of 10 g or more. For these sections, conductors could be located in the thruster slab to provide strong vertical or lateral magnetic forces to offset the centripetal forces on the spacecraft.

FIGURE 3.18A.

CURRENTS AND FORCES FOR STABILITY LOOPS FOR CENTERED AND NON-CENTERED SPACECRAFT

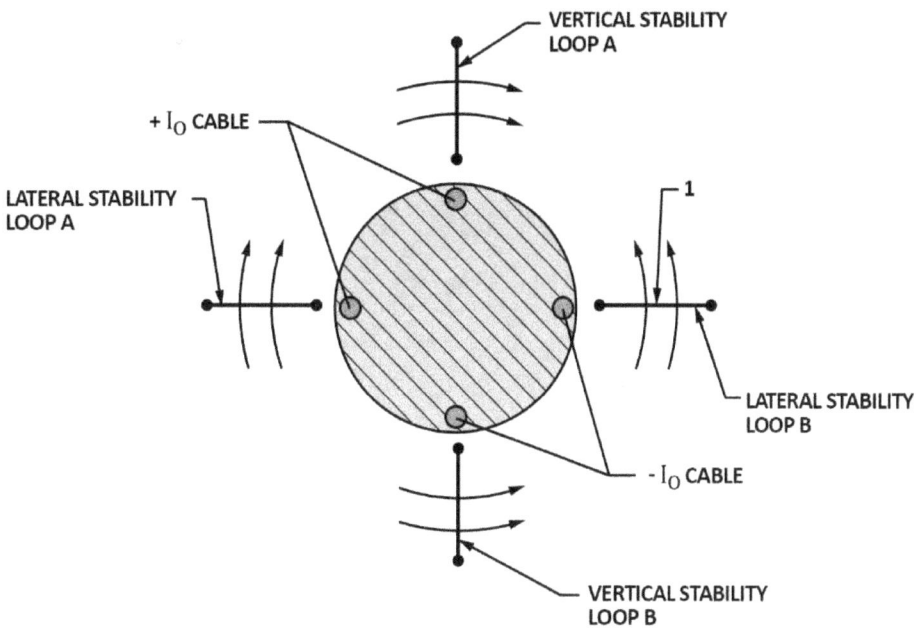

Vertical stability loops A & B connected in null flux circuit
Net flux In circuit is zero – no current flows in vertical stability loops
Lateral stability loops A & B connected in null flux circuit
Net flux In circuit is zero – no current flows in lateral stability loops

FIGURE 3.18B.
SPACECRAFT DISPLACED VERTICALLY TOWARDS LOOP A
–NET FLUX THRU CIRCUIT LOOP CURRENT PUSHES SPACECRAFT BACK TO CENTER

FIGURE 3.18C.
SPACECRAFT DISPLACE VERTICALLY TOWARDS LOOP B
– NET FLUX THRU CIRCUIT LOOP CURRENT PUSHES SPACECRAFT BACK TO CENTER

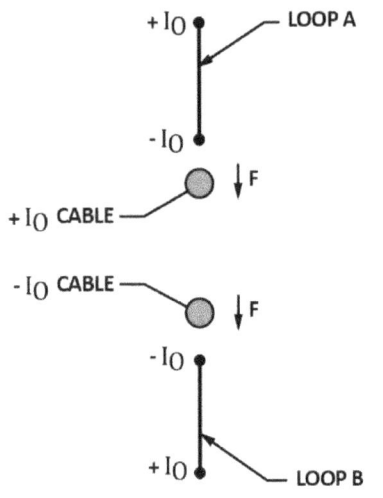
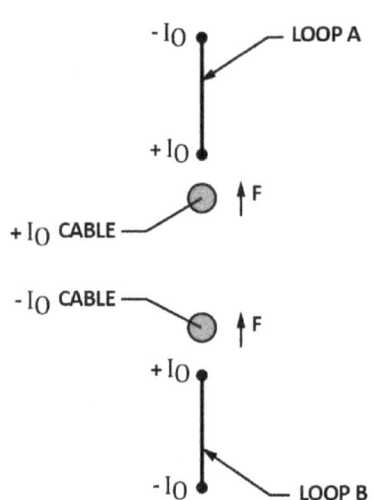

Status of Key Technologies for StarTram

The 8 key technologies for the StarTram system are listed below:

- StarTram Spacecraft levitation and propulsion
- StarTram Spacecraft Superconducting Loops
- MACE Energy Storage
- Tunnel Construction
- Evacuated Launch Tube Construction
- MHD Exit Window
- Atmospheric Drag/testing of StarTram Spacecraft
- Orbital Insert of StarTram Spacecraft

Starting with the first key technology, StarTram Spacecraft levitation and propulsion, it is important to recognize both how it relates to existing Maglev technology, and what new developments are needed to implement StarTram.

Japan's development of their 1st generation passenger Maglev transport system has firmly established the practicality of Maglev levitation and propulsion. Japan's Maglev system uses superconducting loops on the vehicles to interact with null flux aluminum loops on the guideway, the 1966 invention of Powell and Danby, to levitate and automatically strongly stabilize high speed vehicles with large clearances between them and the guideway. The weight of Japan's levitated Maglev vehicles is comparable to that of a StarTram spacecraft and the method of levitation very similar.

The principal difference between Japan's Maglev system and StarTram's Maglev system is the method of propulsion. Japan's system uses the Linear Synchronous Motor (LSM) in which the vehicle is magnetically propelled by the interaction between the DC superconducting magnets on the vehicle with a traveling AC current wave in aluminum loops on the guideway underneath.

The LSM system, also invented by Powell and Danby, is very effective and practical for Maglev vehicles traveling at several hundred miles per hour. However, it does not appear possible for StarTram because of the spacecraft's much greater velocity—8 kilometers per second, (18,000 mph) at exit from the launch tube. The required thruster currents, AC frequencies, and inductive voltages are very high and are not practical at such velocities.

Instead, the pulsed DC current thruster system described in this report appears practical. However, it will require development and testing, initially using low-cost sub-scale modular experimental units to establish feasibility. There is no need to build a full-scale system from scratch before its feasibility is established. Once pulsed DC thruster system is proven by modular operation, the full scale StarTram launch system can be built.

The next key technology, the StarTram Spacecraft Superconducting Loops, is a new application of existing superconductor technology. Japan's 1st generation superconductor operates with liquid helium coolant at 4 degrees Kelvin. The superconductor is thermally insulated, with liquid helium refrigerators maintaining the superconductor at 4 degrees Kelvin.

The StarTram spacecraft will use a new, higher performance high temperature superconductor (HTS) based on very thin, approximately 1 micron thick, film of Yttrium Barium Copper Oxide (YBCO) deposited on thin steel/copper tape. HTS superconductor is commercially available with very high field, very high current density.

The StarTram spacecraft has 2 superconducting cable loops imbedded in the Sabot that encloses the actual payload, as illustrated in Figure 3.5. The loops are 5 meters in length along the length of the spacecraft, as illustrated in Figure 3.6. The cables have a thin layer of thermal insulation around their surface, but remain at ambient temperature until just before launch.

A few minutes prior to launch, the cables would be cooled down to 4 degrees Kelvin and energized with current. The spacecraft is then accelerated to 8 km/sec in 20 seconds. There is no-onboard refrigeration on the space craft – the cables have sufficient thermal inertia to maintain their 4-degree temperature during the acceleration period.

The total length of the superconducting cables on the StarTram spacecraft is approximately 30 meters, about 1/2000th of the total superconducting cable length in the 27-km circumference of the Large Hadron Collider (LHC) particle accelerator that operates in CERN, Switzerland.

Cooling down the spacecraft's superconducting cables 5 minutes prior to launch time, they would only be at 4 degrees Kelvin for 1/100,000th of a year, whereas the CERN superconducting cables operate annually at even lower temperature, 1.7 degrees Kelvin, with much greater refrigeration power.

Even if one were to launch 1,000 StarTram spacecraft per year, the refrigeration cost would be completely negligible compared to the refrigeration cost for the CERN superconducting cables. Total yearly energy cost for the LHC, refrigeration plus equipment plus lights, etc. is 19 million Euros, about 25 million dollars. StarTram spacecraft cable refrigeration costs will be tiny by comparison, e.g., at most $100,000 per year. At 1,000 launches per year and 20 tonnes payload per launch, that's only $100/29,000 kg or 5 cents per kilogram of payload.

Present costs of HTS Superconductor are on the order of $50 per kiloamp meter. The 20 tonne payload StarTram spacecraft uses approximately 120,000 kiloamp meters of HTS Superconductor. At $50 per kiloamp meter that corresponds to a cost of 6 million dollars per launch, or about $300 per kilogram of payload.

HTS superconductor cost is rapidly declining, however, like Nb-Ti superconductor has done. In its early days Nb-Ti superconductor cost $50 per kiloamp meter, but is now at $1 per kiloamp meter.

With large scale production HTS superconductor prices should drop to $5 per kiloamp meter, and probably even less.

At $5 per kiloamp meter, StarTram superconducting cable would cost $600,000 or $30 per kilogram of payload. At $2 per kiloamp meter, which should be achievable with large scale production, superconductor cost would be $12 per kilogram of payload.

The third key technology, MACE Energy Storage builds on the R&D on Superconducting Magnetic Energy Systems (SMES) work now going on in connection with energy storage for utilities and pulsed power for weapons systems, e.g., railguns. Small SMES units have been built and tested and appear feasible. Experimental pulsed power systems based on inductively transferring magnetic energy from a superconducting coil to a closely coupled ordinary copper coil have also been built and tested, demonstrating concept feasibility.

The MACE design described have is an evolved version of earlier SMES and pulsed power system with greater energy storage capacity and increased energy transfer efficiency. It has been proposed for testing, but not yet funded.

The StarTram system is based on multiple identical modular units, e.g., 800, which can be mass produced at acceptable cost and readily installed along the 80-kilometer evacuated launch tube. With appropriate redundancy and rapid replacement capability, StarTram can continue to operate even if individual MACE units were to fail.

The 4th key technology, Tunnel Construction, is well within present commercial technology. There are many long and large tunnels throughout the world that demonstrate the capability to construct the 80 kilometer StarTram tunnel.

Table 3.4 lists some of the long tunnels in the World, (3) their location, length, diameter and excavation volume. The Large Hadron Collider (LHC) tunnel in CERN, Switzerland is 27 kilometers in circumference. The Superconducting Super Collider (SSC), which was to be constructed in Texas was cancelled after 22.5 kilometers of its planned 87-kilometer tunnel had been constructed. There are lots of railway and water supply tunnels that have been constructed that are 50 kilometers or more in length.

Figure 3.19 shows a schematic view of the 3 parallel StarTram Launch tunnels and their function. The tunnel at the right side of Figure 3.19 contains the 80-kilometer evacuated launch tube. Beneath the launch tube is an excavated portion with a conventional railway track that transports prefabricated 100-meter launch tube Sections on 3 coupled conventional flatbed railway cars. When the appropriate location is reached the 100-meter Section is raised by hydraulic lifter and anchored to the tunnel structure.

This minimizes field construction and installation requirements. The launch tube Sections are shipped fully fabricated from the factory to the construction site by railway with all their thruster cables, stability loops, sensors, etc. already attached. When lifted into place, the Section is vacuum sealed to its adjacent Section and connected to its MACE energy supply unit.

TABLE 3.4.
LONG TUNNELS IN THE WORLD

Type/Name	Location	Length	Diameter	Excavation Volume
Water Supply				
Delaware Aqueduct	New York	137 km	4.1 m	$1.8 \times 10^6 m^3$
Päijänne Tunnel	Finland	120 km	4.5 m	$1.9 \times 10^6 m^3$
Dahuofang Tunnel	China	85 km	8.0 m	$1.9 \times 10^6 m^3$
Orange-Fish River Tunnel	South Africa	83 km	5.4 m	$1.9 \times 10^6 m^3$
Bolmen Tunnel	Sweden	82 km	3.2 m	$1.9 \times 10^6 m^3$
Emisor Oriento	Mexico City	62.4 km	-	-
Zelvika Tunnel	Czech Republic	52 km	2.4 m	$0.3 \times 10^6 m^3$
Railway/Transit Tunnels				
Guangzhou Metro	China	60.4 km	-	-
Beijing Subway	China	57 km	-	-
Gotthard Base	Switzerland	57 km	9 m (2 tubes)	$8.1 \times 10^6 m^3$
Seikan Tunnel	Japan	54 km	9.5 m	$3.8 \times 10^6 m^3$
Channel Tunnel	England/France	50 km	7.6 m (2 tubes) 4.6 m (1 tube)	$5.5 \times 10^6 m^3$
Particle Accelerator Tunnels				
Large Hadron Collider	CERN, Switzerland	27 km	3.8 m	$0.3 \times 10^6 m^3$
Superconducting Super Collider*	Texas	87 km	4.2 m	$1.2 \times 10^6 m^3$

* Cancelled in 1993 after 22.5 kilometers of the 87 kilometers long tunnel had been excavated.

The smaller service tunnel, shown in the middle of Figure 3.19 provides access for personnel to inspect and maintain the evacuated launch tube in the adjacent tunnel, as well as the MACE energy storage unit located in the second adjacent tunnel. The MACE unit illustrated in Figure 3.19 is a linear dipole loop of superconducting cable, rather than the circular MACE loop described earlier. Either geometry appears suitable. The linear dipole MACE loop uses a longer length of superconducting cable, but will perform as well as the circular MACE loop.

The recently completed Gotthard Tunnel through the Swiss Alps (Table 3.4), the longest railway tunnel in the world, (4) has 2 railway tubes, each 57 kilometers in length and 9 meters in diameter. Total excavation volume was 8.1 million cubic meters.

Total construction cost for the project, excavation plus track, rail equipment, stations, etc. was 10 Billion dollars, corresponding to a unit cost of approximately $1200 per cubic meter of excavated tunnel.

Table 3.5 given the dimensions and excavation volume for the 3 StarTram tunnels illustrated in Figure 3.19. All 3 tunnels run the full 80-kilometer length of the StarTram launch facility. Total excavation volume is 7.2 million cubic meters, less than the 8.1 million cubic meters for the Gotthard Tunnel, and somewhat greater than the 5.5 million cubic meter excavation for the 3 tunnels in the 50-kilometer-long Chunnel, the railway tunnel between England and France.

FIGURE 3.19.

SCHEMATIC VIEW OF THE STARTRAM LAUNCH TUNNELS AND THEIR FUNCTIONS

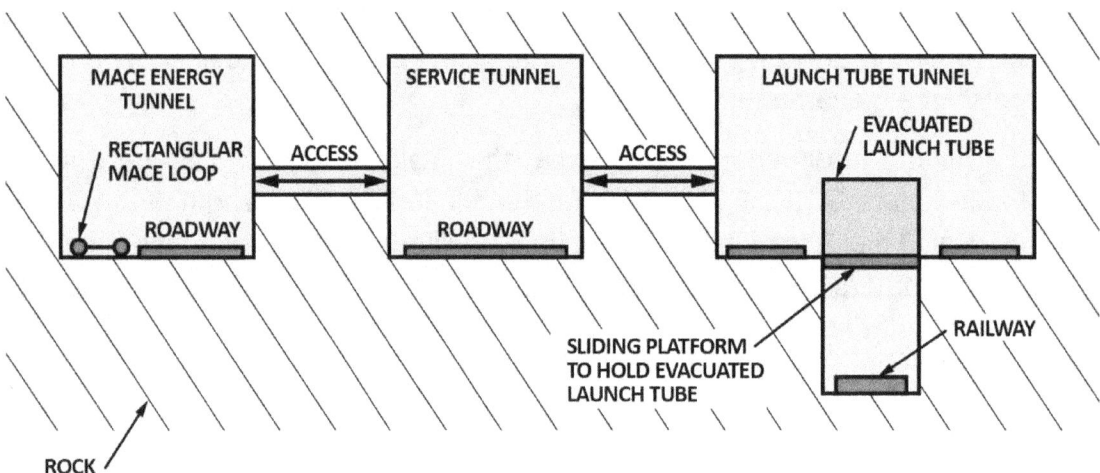

Taking the unit cost of excavation, etc., for the StarTram Tunnels to be the same as that for the Gotthard Tunnels, i.e., $1,200 per cubic meter, the total construction cost of the StarTram Tunnel is $8.6 Billion, a bit less than the $10 Billion for the Gotthard Tunnels.

Conclusion. The 3 StarTram tunnels can be constructed using current technology at acceptable cost, and will be less expensive than the Gotthard Railway Tunnel through the Swiss Alps.

The 5th key technology, Evacuated Launch Tube Construction is also available today. The Large Hadron Collider (LHC) at CERN in Switzerland uses a 0.9-meter diameter high vacuum cylinder to contain the LHC cryogenic magnets and high energy particles beam pipes. The total circumference of the LHC 0.9 meter vacuum vessel is 27 kilometers.(5)

The pressure inside the LHC vacuum vessel is on the order of 10^{-9} atmospheres, far lower than will be required for the evacuated StarTram launch tube. For StarTram, a passive level of 10^{-6} atmospheres is more than adequate, and readily available using existing vacuum and cryopump technology.

The Large Hadron Collider structure inside its 0.9 meter diameter, 27 km long vacuum vessel is far more complex that the structure of the StarTram evacuated launch tube. The LHC has very complex superconducting magnets that operate at 1.9 degrees Kelvin that have to be positioned extremely precisely with very precise magnetic fields that ramp up in time. Also, it has complex and very precise pipes which carry the very high energy proton beams.

TABLE 3.5.
CONSTRUCTION COST OF 3 STARTRAM TUNNELS

Basis:
80-kilometer length
$1,200 per m3 Excavation Cost

Tunnel Type	Tunnel Diameter	Volume	Construction Cost
#1 Launch Tube Tunnel	5x8 meters	$3.2 \times 10^6 m^3$	3.8 B$
#2 Service Tunnel	5x5 meters	$2 \times 10^6 m^3$	2.4 B$
#3 MACE Energy Storage Tunnel	5x5 meters	$2 \times 10^6 m^3$	2.4 B$
	Total	$7.2 \times 10^6 m^3$	8.6 B$

Figure 3.20 shows a cross section of the LHC vacuum vessel and its internal structure. There are 1200 separate beam magnets in the LHC, each 15 meters in length, all which have to be installed precisely, with many hundreds of joints required for currents, controls sensors, and vacuum vessel connections.

The StarTram evacuated launch tube is much simpler. 100-meter prefabricated launch tube sections that have been fully tested and validated with all of their equipment attached are shipped to the construction site and joined to adjacent sections with a simple straightforward vacuum seal between the sections.

The proposed "Hyperloop" transport evacuated tube system between Los Angeles and San Francisco, a distance of 670 kilometers (400 miles) a length almost 10 times greater than the StarTram launch tube. Moreover, the length of the "Hyperloop" evacuated tube sections is considerably smaller than the 100-meter StarTram Sections, so that the number of sealings of vacuum sections needed for Hyperloop would be more than 30 times greater than for StarTram.

Conclusion. Construction and installation of the StarTram 80 km evacuated launch tube will be much simpler and easier than the construction already carried out for the 27 km Large Hadron Collider's vacuum vessel. It will also be simpler and easier than the evacuated tube proposed for the "Hyperloop" transport system.

The sixth key technology, the MHD Exit Window keeps the outside atmosphere from entering the evacuated launch tube as the StarTram spacecraft is accelerating along its 80-km length. During the period prior to a spacecraft launch, the evacuated launch tube is closed off at its exit end by a primary mechanical shutter that prevents outside air from entering the launch tube.

Immediately prior to the start of the launch, the primary shutter is opened, and the spacecraft acceleration initiated. An explosively driven secondary shutter remains closed for most of the 20 second acceleration process, until the spacecraft is only a few seconds from the exit. It then is driven open, removing any mechanical blockage of the exit.

Conventional steam jet ejectors located around the launch tube exit begin operation before the spacecraft begins to accelerate, directing jets of high pressure steam outwards from the exit region. The stream jets reduce the air density at the launch tube exits to approximately 1/10th of normal air density, i.e., to about 0.1 kg/cubic meter, equivalent to the atmospheric density at an altitude of 20 kilometers.

To prevent this low-density air from flowing into the evacuated launch tube, the final section of the launch tube has an "MHD window", consisting of a strong traverse magnetic field across the tube supplied from a set of superconducting cables located outside of the tube.

Outside air flowing into the tube that encounters the transverse magnetic field is ionized by RF discharge. DC voltage is also applied across the tube, causing current to flow through the ionized air. The product of the DC current in the ionized air creates an outward hydrodynamic force on the air toward the exit, and prevents it from flowing into the launch tube.

FIGURE 3.20.
LHC VACUUM VESSEL CROSS SECTION

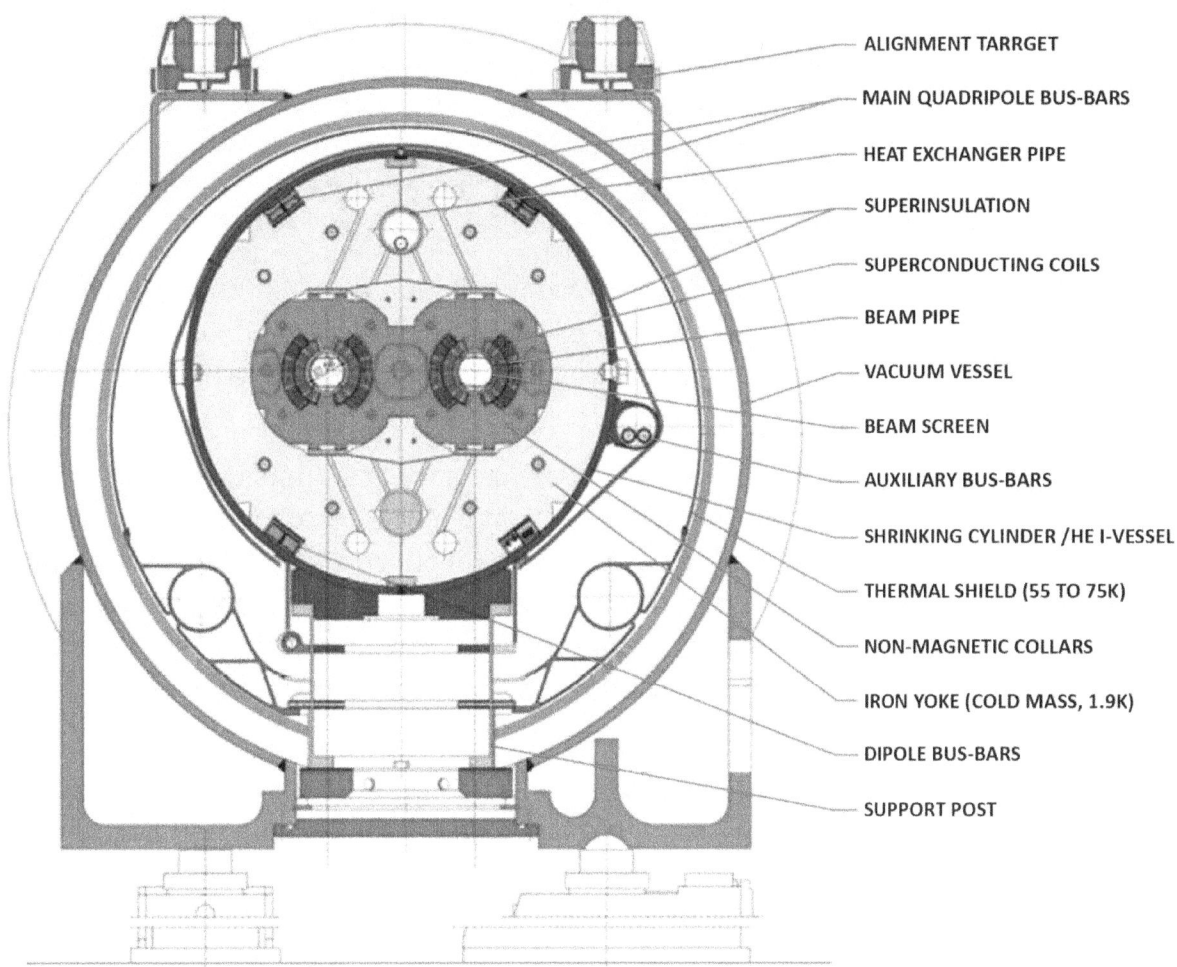

This effect of a magnetic field and current hydrodynamically pushing on ionized air is the same as that in Magneto Hydro Dynamic (MHD) generators, with the difference that the MHD window acts as a pump on the ionized air using an applied DC current and applied power, while in MHD generators, flowing ionized air through a transverse magnetic field generates DC current and power.

MHD generators are a well-established technology, with MHD units that have generated hundreds of megawatts of electric power. Their technology can be readily applied for the MHD window system. While technically successful, MHD generators have not been economically successful for electric utilities, because they are more expensive than gas turbines. However, the cost of the MHD window for StarTram use is very acceptable.

Conclusion. The combination of mechanical shutter, steam jets, and the MHD window will prevent significant amounts of outside air from entering the StarTram launch Tube when the spacecraft is launched. There will be a small amount of leakage which can be removed using mechanical vacuum pumps and cryopumps.

The seventh key technology, Atmospheric Drag and Heating of the Spacecraft is clearly related to the technology used in re-entry vehicles launched on ICBMs. The re-entry vehicles re-enter the atmosphere at velocities close to the 8 km per second StarTram spacecraft launch velocity, traveling through even higher atmospheric densities than the StarTram spacecraft.

StarTram spacecraft launched from a high altitude, e.g. 6,000 meters, facility experience an initial atmospheric density of about 45% of the atmospheric density at sea level, where most of the re-entry vehicle targets are located. The heating and drag forces on the StarTram spacecraft will be substantially less than on a re-entry vehicle.

Furthermore, the ratio of surface area to body mass for StarTram spacecraft is much smaller than that for re-entry vehicles, enabling much more mass for protection against aerodynamic heating, including transpiration cooling, which is extremely effective in preventing excessive surface temperature. The lower rate of surface area to mass for StarTram compared to re-entry vehicles also considerably reduces the ΔV velocity decrease due to atmospheric drag, compared to re-entry vehicles.

As the Gen-1 cargo craft exits the acceleration tunnel and begins its climb from ground level to orbit, it experiences strong aerodynamic heating and deceleration forces. For a given cargo craft weight and size, the magnitudes of the heating and deceleration forces depend on three factors: 1) altitude of atmospheric entry, 2) nose geometry of the cargo craft, and 3) launch velocity.

At a launch velocity of 8 km/sec and a launch altitude of 8000 meters, a blunt nose results in a deceleration rate of 12g, while a sharp nose results in a deceleration rate of 6g. The peak heating rate at the stagnation point is 20 KW/cm^2 for the blunt nose, and 100 KW/cm^2 for the sharp nose. The deceleration and heating are higher at lower launch altitudes, because of the differences in atmospheric density. At 4000 meters launch altitude, for example, the ratio of atmospheric densities, $\rho(4000)/\rho(8000)$ is 1.56, resulting in a proportionate increase in deceleration rate. The percentage increase in heating rate is somewhat smaller, about 30%, for the 4000 m altitude launch. The deceleration and heating are higher at lower launch altitudes, because of the greater atmospheric density is approximately 50% greater, resulting in a proportionate increase in the aerodynamic deceleration force, but still acceptable.

These deceleration and heating rates are less than those encountered when reentry vehicles carrying weapon payloads travel downward through the atmosphere. Moreover, the Gen-1 spacecraft can carry large amounts of water for transpiration cooling of high heat areas, a very effective way of handling the heat loads. Pertinent to this issue, the German Aerospace Center is expending considerable effort, theoretical and experimental, on developing suitable porous, ceramic materials for transpiration water cooling of the nose and leading edges of the Space Liner, FAST 20XX. Because of the high payload efficiency and very low cost of Maglev Launch, this does not significantly affect performance and cost. In comparison, reentry vehicles are much more limited in transpiration cooling capability.

The cargo craft will lose some ΔV (change in velocity) as it travels upward through the atmosphere to orbit because of aerodynamic drag. The magnitude of the ΔV loss will depend on four factors: 1) launch altitude, 2) launch angle, 3) drag coefficient, and 4) launch velocity.

In order to coast to the desired orbital altitude, the cargo craft needs the necessary velocity as it leaves Earth's atmosphere. For insertion into LEO, the required velocity is ~8 km/sec. Based on detailed design studies, three conclusions can be drawn:

Launch altitudes as low as 4000 meters (13,000 feet) are practical. Higher altitudes are desirable, because the deceleration, heating rate, and ΔV loss through the atmosphere is less. Launching at 8000 meter altitude, for example, reduces the ΔV loss through the atmosphere by almost a factor of 2.

The optimum launch angle will probably be in the range of 10 to 15 degrees. Both result in acceptable ΔV loss, though 10 degrees yields a greater value for ΔV. Less than 10 degrees would result in a larger ΔV loss, while much greater than 15 degrees would require a large ΔV burn for orbit insertion.

The simplest, cheapest, and easiest way to compensate for the ΔV loss through the atmosphere is to accelerate the cargo craft to $(8+\Delta V)$ km/sec in the StarTram acceleration tunnel. The additional ΔV acceleration requirement is modest on the order of 0.5 km per second, and eliminates the need for a compensating rocket motor on the cargo craft. (A small rocket is still needed to circularize the orbit.)

Where would the Gen-1 and Gen-2 launch sites be located? For Gen-1, potential sites should meet the following criteria:

- Launch altitudes of 4000 meters or more.
- Remote location with very low population density.
- Minimum flight length over land.
- Launch into polar orbits.

Criteria #1 keeps aerodynamic drag and heating at practical levels, and minimizes the ΔV loss during ascent through the atmosphere. Criteria #2 avoids disturbing people with very intense sonic booms. Criteria #3 minimizes hazards to population from debris if a cargo craft failed during its ascent to orbit.

Criteria #4 reflects two factors—first, for high resolution environmental monitoring and broadband communications polar LEO orbits are desirable because they cover all areas on Earth. Second, potential sites for equatorial or low latitude orbit launches that meet criteria 1, 2, and 3 are much fewer than those that can launch into polar orbits. The Moon and Mars colony applications and robotic space exploration missions could use either polar orbits or equatorial orbits. Transferring assemblies of SPS satellites from polar orbit to GEO equatorial orbit would require a ΔV change. Because of the low launch cost of the propellant needed for the orbit transfer using Gen-1 and the high Isp.(specific impulse of the thruster for orbit transfer) of the solar powered thruster, the orbital transfer is low cost.

A launch site location that meets all four criteria is Antarctica. The Vinson Massif in Marie Byrd Land reaches 5,140 meters altitude. From there a launch vehicle would fly over the open Pacific to Alaska—halfway around the world—before passing over land. By then it would already be in orbit or destroyed over open water if some defect developed. There is no local population and no problem of sonic booms. The only drawback to Antarctica is its isolation.

This should not be a problem. Large container ships already carry massive amounts of freight halfway around the world at low cost. Four other possible isolated sites for polar launch are shown in Table 3.6. The sites are located in host countries having substantial space programs. Potential sites in the Himalayas and Andes could also launch at very high altitudes, i.e. 6,000 to 8,000 meters; however, population densities are higher.

TABLE 3.6.
FEATURES OF POTENTIAL SITES FOR A GEN-1 STARTRAM FACILITY.

Feature	United States	Russia	China	Greenland (Europe Host)
Name/location of mountain peak	Mt. St. Elias, Alaska	Klyudevskaya Sopka, Kamchatka Peninsula	Gongga Shan Szechwan Province	Highest point in Greenland ice sheet
Altitude of peak, meters	5489	4750	7556	3220
Altitude of exit point from acceleration tunnel, meters	4989	4250	7050	3220
Launch angle (LA), degrees	10	10	10 < LA < 15	10
Flight distance to reach first land flyover beyond host country	15,000 km flight over Pacific Ocean	14,000 km flight over Pacific Ocean	600 km flight over China	18,000 km flight over Atlantic Ocean
Location of first land flyover	Rockefeller Plateau Antarctica	Wilkes Land, Antarctica	Vietnam	Filchner Ice Shelf, Antarctica
Notes:	Second land flyover is 13,000 km farther downrange (Africa)	Second land flyover is 20,000 km further downrange (Greenland)	Second land flyover is 3000 km further downrange (Indonesia)	Second land flyover is 5000 km further downrange (Australia)

Three of the Gen-1 sites—Alaska, Greenland, and Kamchatka—have very long downrange flight distances over the open ocean, i.e. 14,000 km or more, before reaching Antarctica, an unpopulated continent; after passing over Antarctica they have 1000s of additional flight miles over the open ocean before reaching a second land mass. For the Kamchatka site, the second land mass is Greenland, which is essentially unpopulated, with a flight distance of 20,000 kilometers before it is reached.

A complete description of the drag and heating effects on the spacecraft as it ascends through the atmosphere to orbit is given in StarTram: The New Race to Space by Powell, Maise, and Pellegrino, available at Amazon.com

Conclusion. Protecting the surface of StarTram Spacecraft from excessive atmospheric heating and minimizing ΔV losses due to atmospheric drag forces will be much easier than for re-entry vehicles, which can transmit the atmosphere successfully and function when they reach their targets.

The 8th key technology, Orbital Insert of StarTram Spacecraft is well established and has been applied many times to insert satellites launched by chemical rockets into their final orbits. A small ΔV burn by a chemical rocket establishes the final orbit.

For the StarTram spacecraft, insertion with LEO (Low Earth Orbit) requires a ΔV burn of 0.34 km/sec, about 4% of the launch velocity of 8 km/sec. Insertion into GEO orbit 22,00 miles above Earth requires a larger ΔV burn, 1.5 km/sec, about 20% of its launch velocity. The very low launch cost per kilogram for StarTram makes the cost of the rocket for orbital insertion small and acceptable.

Conclusion. Orbital insertion of StarTram Spacecraft will be very similar to that already carried out using chemical rocket technology and will not require new technology.

Construction and Operating Costs of the StarTram Launch Facility

In this section the construction and operating costs of the StarTram launch facility are described.

The principal capital cost components of the StarTram Facility are the:

Launch Tube Tunnels

Thruster Circuits and Slabs

MACE Energy Storage Units

Evacuated Launch Tube

The costs projected here are based on the 80-kilometer-long launch tube design described previously that launches 28 tonne spacecraft carrying 20 tonnes of payload at 8 km/second.

Starting with #1, the Launch Tube Tunnels, as describe earlier, the estimated construction cost is 8.6 Billion dollars, based on the average excavation cost of $1200 per cubic meter of volume for the recently completed 57 kilometer Gotthard tunnel through the Swiss Alps. The $1200/cubic meter is derived from the total cost of the Gotthard railway tunnel and the total excavation volume. It includes not only the excavation cost but also all the equipment installed in the tunnel. The excavation cost for the StarTram facility will probably be somewhat less than $1200/cubic meter if the Gotthard equipment cost were not included, so that the StarTram projected tunnel construction cost is probably an over estimate.

The next main component is the thruster slabs, their cables and conductor shields. Table 3.7 lists the materials used and their amounts for a 100-meter-long thruster slab. There are 4 thruster slabs for each 100-meter-long launch tube section, and 800 sections in the 80-kilometer-long launch tube, making a total of 3200 thruster slabs for the StarTram launch facility.

The thruster slabs would probably be manufactured in shorter lengths than 100 meters, then joined together to form a 100-meter slab, to which the thruster cable circuit would be attached. The finished 100-meter slab would then be joined for the 100-meter launch tube section and transportation to the assembly site using the rail track in the StarTram tunnel.

The estimated costs per tonne of copper conductor and epoxy fiberglass are taken to be 3 times the market cost of $4,500 per tonne for copper, and 3 times the market cost of epoxy fiberglass at $4,000 per tonne, reflecting the additional cost for their fabrication and installation into the thruster cables and slabs.

The total capital cost of the 3200 thruster cable slabs – 4 per 100-meter launch tube Section and 800 launch tube Sections for the 80-kilometer StarTram launch tube system is 4.6B$, about ½ of the tunnel cost of $8.6 Billion.

Using aluminum conductor instead of copper conductor would significantly reduce the conductor cost for the thruster slabs. More detailed analysis can determine the potential savings. The temperature rise in aluminum conductor will be somewhat greater but appears acceptable.

TABLE 3.7. COST OF STARTRAM THRUSTER SLABS

Basis: 100-meter length
4 slabs per 100-meter launch tube section, 800 sections
140 meters of copper conductor shielding, 2 cm this for 80 km launch tube, 2 cm ID
280 meters of thruster copper cable
200 cm2 thruster cable cross sectional area
30 cm thick, 100-meter total length, 1.2-meter-wide thruster slab, epoxy fiberglass

Material	Amount/100 Meter	Fabricated cost/tonne	Total Cost for 3800 Slabs
Copper Thruster Cables per Slab	50 tonnes	$13,500/tonne	2.2 B$
Copper Conductor Shields per Slab	16 tonnes	$13,500/tonne	0.69 B$
Expoxy Fiberglass Slab	45 tonnes	$13,000/tonne	1.7 B$
		Total	4.6 B$

Conclusion. The cost of the thruster slabs appears reasonable and less than the cost of the StarTram tunnel system. More detailed analysis is required for a more precise estimate of thruster slab cost, but it appears to be a small part of total capital cost. Amortized over a 30-year operating life of a StarTram facility that launches 100,000 tonnes of payload per year, its 4.6 B$ capital cost is only $1.50 per kilogram of payload launched.

The third main component of capital cost is the 800 MACE energy storage units. Table 3.8 gives the capital cost of a MACE energy storage unit, and Table 3.9, its operating cost, principally the electric power for refrigeration.

Total capital cost for the 800 MACE units is 4.8 Billion$, with most of the capital cost from the High Temperature Superconductor. The $5 per kiloamp meter is 1/10th of the current small-scale production cost of approximately 50$ per kiloamp meter for YBCO HTS tape superconductor.

With the larger market expected for HTS Superconductor, a major reduction in unit cost appears very likely. Chinese suppliers indicate, for example, that $20 per kilo amp meter appears possible in the near term for large orders. The cost of the material for HTS superconductors tape is extremely small. One kilo amp meter of YBCO tape uses less than 1 milligram of YBCO material deposited on stainless/copper tape that weighs less than 10 grams. Materials cost only a few cents. Today, processing using expensive machines is the primary cost. With large-scale, automated machine production, cost will be much less in the future.

TABLE 3.8.
CAPITAL COST OF MACE ENERGY STORAGE UNIT

Basis:
3.3 Gigajoules energy storage per MACE unit
800 MACE units for 80 km StarTram Launch System
30 degrees Kelvin Primary HTS Superconducting Winding
10 million amp turns of HTS Superconductor in Primary Winding
MACE circular loop geometry, 27.5-meter diameter
25-centimeter radius of superconducting HTS winding
850,000 kiloamp turns of HTS superconductor
5-centimeter-thick multilayer vacuum thermal insulation
1×10^{-4} W/mK thermal conductivity of Multilayer Thermal Insulation
5-centimeter-thick secondary copper winding @ 300K
40 centimeter out radius of MACE cable (0.8 m OD)

Material	Amount Per MACE Unit	Unit Cost	Cost for a MACE Unit	Cost For 800 MACE Units
HTS Superconductor	830,000	$5/KAM	4.2 M$	3.4B$
Multi-Layer Thermal Insulation	137 m2	$2,000/m2	0.3M$	0.3B$
Epoxy Fiberglass	11 tonnes	$12,000/tonne	0.11M$	0.1B$

Support Tube				
Secondary Copper Winding	90 tonnes	$13,500/tonne	1.2M$	1.0B$
		Total	5.8 M$	4.8 B$

Amortized Capital Cost for MACE Units Per Kg of Payload

Basis: 100,000 tonnes of payload/year, 30 years, 4.8B$ Capital Cost.

Cost/kg = $\frac{4.8 \times 10^9}{10^5 \times 10^3 \times 30}$ = $1.6/kilogram

Amortized over a 30-year period, with 100,000 tonnes of payload launched annually from the 80 Km StarTram launch facility, the amortized capital cost of 800 MACE units is only $1.6 per kilogram of payload, a very small amount (Table 3.8).

Even if the cost of HTS superconductor were to be greater than expected, its effect on cost per Kg of payload launched would be minor.

The operating cost per Kg launched for refrigeration is even smaller than the amortized capital cost of the MACE unit.

Table 3.9 shows the cost components for refrigeration of the 30K HTS superconductor winding. The HTS winding is thermally insulated from the 300K epoxy fiberglass support tube around which the secondary copper winding is wrapped. The insulation layer is Multi-Layer Insulation (MLI), consisting of multiple layers of aluminum foil and fiberglass sheets in a high vacuum enclosure.

MLI insulation is standard for cryogenic applications and widely used throughout the world. It has a thermal conductivity of 1 x 10^{-4} W/mK. The effective thermal conductively will be greater, however, because of the need to mechanically support the weight of the primary HTS winding inside the epoxy fiberglass support tube, using low thermal conducting composite supports – e.g. fiberglass epoxy.

Adding in the supports, the effective thermal conductivity of the MLI insulation plus supports is projected to be 5 x 10^{-4} W/mK, 5 times the thermal conductivity of the MLI insulation.

With a 270 K temperature difference between the 300K support tube and the 30K HTS winding and a 5-centimeter-thick insulation layer, the thermal leakage into the HTS winding is 2.7 watts (th) per square meter of insulation area, for a thermal conductivity of 5 x 10^{-4} w/mK.

The thermal leakage into the HTS winding is removed by a cryocooler that uses electricity to power a refrigeration cycle. Cryocoolers are commercially available that operate at 30 degrees Kelvin. The refrigeration factor for removing thermal energy from a 30K region, watts (e) per watts (th), is on the order of 20% of a Carnot cycle that operates between 270K and 30K, with a value of 45 watts (e) per watt (th). The corresponding electric refrigerator power per square meter of insulation area is then 45 watts(e) x 2.7 watt(th)/m² or 120 watts(e)/m².

For 1 MACE unit with 137 m² of insulation area, the refrigeration electric power is 16 kW(e), and for 800 MACE units for the StarTram launch System, a total of 13 megawatts(e). The cost of the refrigeration power at 10 cents per KWH per kg of payload launched at 100, 000 tonnes per year is only $0.11 per Kg of payload, a very small amount.

TABLE 3.9.
OPERATING COST OF MACE ENERGY STORAGE UNITS

Basis:
800 operating MACE units for 80 Km StarTram Launch System
30 Degrees Kelvin HTS Superconducting Winding
25 Centimeter Radius HTS Winding
27.5 Meters Diameter MACE Circular Loop Geometry
5-Centimeter-Thick Multilayer Vacuum Thermal Insulationto
5×10^{-4} W/mK Thermal Conductivity of MLI Insulation
270 K Temperature Difference Across Insulation (300K to 30K)
137 M² Thermal Insulation Area per MACE Unit
2.7 Watt(th)/M² Thermal Leakage Through Insulation Layer
45 Watts(e)/ Watt(th) Refrigeration Factor (20% of Carnot)
120 Watts(e)/M² Refrigeration Power for HTS Winding
16 KW(e) Refrigeration Power for 1 MACE Unit
13 MW(e) Total Refrigeration Power for 800 MACE Units
$215/watt (th) Capital Cost for 30 Degrees Kelvin Cryocooler
$80,000 for Cryocoolers Capital Cost per MACE Unit.

Operating Cost Per Kg furnished for MACE Refrigerators Basis: 100,000 tonnes of payload/year, 30 years' operation Power Cost per Kg of Payload 13 MW(e) x 8760 hours x $100/MWH (10 Cents/KWH) = 11.4 M$/year Cost Per Kilogram of Payload = 11.4 x 10⁶/10⁸ Kg/year =$0.11/Kg
Refrigerator Capital Cost per Kg of Payload Total Capital Cost of Refrigerators = 80,000 x 800 Units = 64 M$ Cost Per Kilogram of Payload = $\dfrac{64 \times 10^6}{10^8 \frac{kg}{year} \times 30\ years}$ = $0.02/kg

The amortized cost of the refrigeration cryocooler equipment is even less. Commercial cryocoolers costing $15,000 can supply 70 watts (th) of refrigeration at 30K, a unit cost of $215 per watt(th). Per MACE unit this corresponds to a cryocooler cost of $80,000, with 64 million dollars for 800 MACE units. Amortized over 30 years at 100,000 tonnes per year of payload, the refrigeration equipment cost is only $0.02 per kg of payload, a trivial amount. If the equipment cost were 5 to 10 times greater, a very unlikely situation, it still wouldn't matter to the cost per kg of payload.

Conclusion. The capital and operating costs of the MACE energy storage units are small in terms of dollars per kilogram of payload launched. The total MACE cost is only $1.73 per Kg of payload.

The 4th main component of the capital cost for the StarTram facility is the cost of the 80 km long evacuated launch tube. Figure 3.7 shows the cross-section of the evacuated launch tube with its thruster. Figure 3.8 show the layout of the 800 launch tube sections, each 100 meters in length, joined together to form the 80-kilometer-long launch facility.

The 100-meter-long evacuated launch tube Sections are cylindrical in shape. The tube has 4 positions spaced 90 degrees apart on its circumference where the thruster slabs are inserted and locked into place. The thruster labs do not penetrate fully through the wall of the tube, so a vacuum seal is not necessary.

The evacuated launch tube has an inner diameter of 2.0 meters, and a wall thickness of 20 centimeters. The stress in the wall due to the outside air at 1 atmosphere is very low with

Swall = (10^5 Pa) (2.0 meters) /2 (0.2 meters = 5 x 10^5 Pa (75 psi)

The mechanical stresses on the wall will be higher when the StarTram spacecraft passes and exerts magnetic forces on the thruster cables and shields.

For a thruster cable force of 1.34 x 10^6 Newtons, 1/8th of the 11 x 10^6 Newtons total propulsion force on the StarTram spacecraft and an area of 0.3 m^2 in the slab that supports the thruster cable (1-meter length x 0.3-meter-thick slab), the stress in the slab would be

Slab = 1.34 x 10^6 N/0.3 m^2 = 4.5 x 10^6Pa (680 psi)

Which is much less than the compressive and tensile strength of epoxy-fiberglass, of more than 20,000 psi.

Accordingly, the evacuated launch tube designs appear very capable of handling the stresses they will experience from the outside atmospheric pressure and the magnetic forces when the StarTram Spacecraft passes.

The quantity of epoxy fiberglass in the evacuated launch tube includes the amount in the cylindrical wall plus the side shoulders that support the thruster slabs, minus the amount of the thruster slabs that fit into the wall of the launch tube. The cross-sectional area of the cylindrical wall, with an ID of 2.0 meters and an OD of 2.4 meters (0.2-meter-thick wall) is

Acyl = $\frac{\pi}{4}$ [(2.4)² − (2.0)²] = 1.4 m²

Adding in the shoulders on the wall and deducting the portion of the thruster slab residing in the insert into the launch tube wall, (A_{total} equals 1.6 meters' square). With a 100-meter-long launch Section, total epoxy -fiberglass volume is 100 x 1.6 =160 m³ with a total weight of 200 tonnes. At $12,000 per tonne, the capital cost of one 100-meter-long launch tube Section would be 2.4M$ and 2 Billion dollars for the 800 Sections in the 80-km long StarTram facility.

Table 3.10 summarizes the capital cost and cost per kilogram of payload for the evacuated launch tube. At 100, 000 tonnes of payload per year for 30 years, the cost of the launch tube is only $0.67 per kilogram of payload launched.

Conclusion. The cost of the evacuated launch tube per Kg of payload launched is small, only $0.67/Kg of payload.

TABLE 3.10.
CAPITAL COST OF EVACUATED STARTRAM LAUNCH TUBE

100 Meter Launch Tube Sections
800 Sections for 80 KM StarTram Launch Facility
2.0-meter ID, 2.4 meter OD Cylindrical Launch Tube
Epoxy – Fiberglass composite material
160 M³ Volume of Epoxy Fiberglass per 100 meters
200 Tonnes Weight of 100 Meter Tube
2.4 M$ Capital Cost of 100 Meter Tube @$12,000/tonne
2 Billion$ Capital Cost of 80 Km StarTram Launch Tube
100,000 tonnes per year of Payload, 30 years' operation
$0.67 per Kilogram of payload launched

Table 3.11 summarizes the capital and operating costs of the 4 main components of the StarTram Launch Facility. The amortized capital cost per kg of payload is $6.70, while the operating cost, stored energy in, and refrigeration of the MACE units is $3.73 per Kg, for a total of $10.43 per Kg of payload. This is very small compared to the present launch cost of $5,000 per kilogram to Low Earth Orbit, using chemical rockets.

Not included is the cost of the StarTram spacecraft including its high temperature superconducting cables. The construction, operation and cost of the StarTram Spacecraft is described in a separate report. It will be much less per kilogram if payload launched for the following reasons.

Much simpler, less complex structure. There are no big rocket engines. The spacecraft simply coasts up to orbital altitude where a small rocket burn established its final orbit.

Much greater payload fraction, e.g., 70% for a StarTram Spacecraft, compared to 4 to 5% for a chemical rocket.

Mass production of large numbers of StarTram spacecraft, with lower cost per unit, compared to the few of a kind now produced for chemical rocket launch.

Much simpler and cheaper launch process than present launch pad systems.

Overall conclusion: The StarTram launch facility capital and operating costs are very low compared to present space systems. 20 Billion Dollars for the StarTram launch facility, compared to 170 Billion spent on the International Space Station, and 200 Billion dollars for the Space Shuttle Program. StarTram will launch much greater amounts of payloads to space, at much lower costs than chemical rockets.

TABLE 3.11.
SUMMARY OF PRINCIPAL CAPITAL AND OPERATING COSTS OF STARTRAM LAUNCH FACILITY

Basis:
20 tonne payload in 28 tonne StarTram Spacecraft
8 km/sec launch velocity
40g acceleration, 11 x 106 Newtons Propulsion Force
1.5-meter Diameter Spacecraft
1.8-meter Diameter Sabot with Superconducting cables on Spacecraft
2.0-meter ID launch tube, 2.4 meter O.D.
100,000 tonnes Payload/year for 30 years

Component	Capital Cost	Amortized Capital Cost/Kg Payload
Three 80 Km long Tunnels	8.6 B$	$2.90/Kg
3200 Thruster Slabs	4.6 B$	$1.53/Kg
800 MACE Units	4.8 B$	$1.60/Kg
80 Km Long Evacuated Launch Tube	2.0 B$	$0.67/Kg
Total Capital Cost	20 B$	$6.70/Kg
Operating Cost		
MACE Stored Energy		$3.60/Kg (3.3 GJ per unit, 800 units, 20, 000 Kg)
Refrigeration of MACE Units		$0.13/Kg
Total Cost/Kg of Payload Amortize Capital Plus Operating		$10.43/Kg

Chapter 3 List of References

1. High Speed Transport by Magnetically Levitated Trains, Powell, J, and Danby, G., Paper 66-WARR-5 presented at 1966 Winter ASME Meeting, NYC, NY (Nov. 1966), A 300 mph Magnetically Suspended Train, Mech Eng 89, p. 30-35.
2. Japanese Maglev Train, Japan Railways
3. List of Longest Tunnels, https://en.wikipedia.org/wiki/Lists_of_tunnels
4. Gotthard Base Tunnel, https://en.wikipedia.org/wiki/Gotthard_Base_Tunnel
5. Large Hadron Collider, https://en.wikipedia.org/wiki/Large_Hadron_Collider

Figure Credits:

Figure 3.1 Photo of 1st Generation Japanese Superconducting Maglev Operating at Yamanashi Test Facility
Figure 3.2 View of StarTram Exiting the Launch Tube
Figure 3.3 The Lorentz Force and StarTram Geometry, Author, Powell
Figure 3.4 The Lorentz Force and StarTram Geometry, Author, Powell
Figure 3.5 Magnetic Acceleration of the StarTram Spacecraft, Author, Powell
Figure 3.6 Layout of Thruster Cable Slab and Current with StarTram Spacecraft, Author, Powell
Figure 3.7 Layout of Thruster Cable and Slab Relative to Superconducting Cable on StarTram Spacecraft, Author, Powell
Figure 3.8 Layout of StarTram Evacuated Launch Tube for Spacecraft Acceleration to Launch Velocity, Author, Powell
Figure 3.9 Drawing of MACE Circuar Loop Energy Storage Unit, Author, Powell
Figure 3.10 Superconductiing Primary and Secondary Copper Winding on Mace Cable, Author, Powell
Figure 3.11 Drawing of Superpower YBCO Superconducting Tape, Author, Superpower, Inc.
Figure 3.12 Drawing of Critical Current of Superpower YBCO As Function of Magnetic Field Strength and Temperature, Author, Superpower, Inc.
Figure 3.13 Geometry and Operation of MACE Energy Storage Unit with Thruster Cable Current, Author, Powell
Figure 3.14A Total Current in MACE Primary and Secondary Windings As Function of Time for 100 meter long section of Launch Tube, Author, Powell
Figure 3.14 B Current in Thruster Cable Current as Function of Time for 100 meter ong Section of Launch Tube, Author, Powell
Figure 3.15 A Voltage and Current in 1 Thruster Circuit as Function of Time Till Spacecraft Passes, Author, Powell
Figure 3.15 B Energy in 1 Thruster Circuit as Function of Time Till Spacecraft Passes, Author, Powell
Figure 3.16A Options For Decay Of Current In Thruster Cable Circuit After Space Craft Passes, Author, Powell

Figure 3.16B Temperature Rise In Thruster Cable Circuit For Options 1 And 2, Author, Powell

Figure 3.17 Vertical And Lateral Stability Of StarTram Spacecraft, Author, Powell

Figure 3.18A Currents And Forces For Stability Loops For Centered And Non-Centered Spacecraft

Figure 3.18b Spacecraft Displaced Vertically Towards Loop–Net Flux Thru Circuit Loop Current Pushes Spacecraft Back to Center, Author, Powell

Figure 3.18c Spacecraft Displace Vertically Towards Loop B – Net Flux Thru Circuit Loop Current Pushes Spacecraft Back to Center, Author, Powell

Figure 3.19 Schematic View of The StarTram Launch Tunnels And Their Functions, Author, Powell

Figure 3.20 LHC Vacuum Vessel Cross Section, https://en.wikipedia.org/wiki/Large_Hadron_Collider#/media/File:The_2-in-1_structure_of_the_LHC_dipole_magnets.jpg, E. M. Henley and S. D. Ellis

Chapter 4

Beaming Low-Cost Solar Power to Earth from Space

To shift away from fossil fuels to renewable energy sources, the renewable sources must meet the following criteria:

- Supply the enormous amounts of energy presently generated by fossil fuels, with the capability to increase supply as world population and standard of living increases
- Provide energy at lower cost than fossil fuels can
- Meet energy needs for all locations in the world
- Not damage the environment
- Capable of rapid implementation
- Welcome by the public

Current Renewable options include:

- Wind turbines and Solar farms
- Hydroelectric
- Nuclear
- Biomass

Biomass can be ruled out. Humanity needs 2,500 calories of food energy (3 kilowatt hours) per day. Using virtually all of the agricultural land on Earth. For 7 Billion persons, biomass would supply only 6,000 Trillion watt hours per year, assuming that people didn't eat.

Sounds like a lot of energy, and it is. However, it is just a small fraction of the total primary energy the 7 Billion world population currently uses, and an even smaller fraction of the energy consumption human society will require in the future.

Currently, humanity consumes 155,000 Trillion-watt hours of primary supply energy 25 times greater than the food it consumes.(1) By 2040, the world primary energy requirement is projected to increase to 250, 000 terrawatt hours,(2) 40 times the food energy we grow. Clearly, biomass is not a practical option for large scale renewable energy.

Nuclear is a more practical option, but it has serious constraints that limit it to being a minor contributor for renewable energy. To supply all of the primary energy of 155,000 terawatt hours now consumed in the world would require 20,000 reactors, each of 1000 megawatts capacity.

The construction cost will be tremendous. At $10,000 per Kw(e) a 1000 megawatt reactor costs 10 Billion dollars. 20,000 reactors would have a construction cost of 200 Trillion dollars, 3 times current annual Gross Domestic Product (GDP).

Nuclear energy will cost much more than energy from fossil fuels, which will rule it out. Also, nuclear reactors are generally not welcomed by the pubic, with strong opposition from many groups. Finally, safety and nuclear waste disposal are major problems for nuclear energy. Large scale implementation of thousands of reactors will require processing the nuclear waste to recover fissile fuel and to vitrify and isolate the radioactive waste from the biosphere.

Hydroelectric cannot meet the very large energy demand if humanity shifts away from fossil fuels. Most practical hydroelectric sites are already generating power, and they only supply a small fraction of the world's present energy needs. There are only a few potential new hydroelectric locations left. Implementing them would not significantly reduce fossil fuel consumption. Moreover, hydroelectric dams and their large reservoirs have major effects on the environment and its ecological systems.

Wind Turbines and Solar Farms are practical sources of renewable energy, and are being significantly implemented, but have important limitations, that raise concerns that while they can reduce fossil fuel consumption, they will not be able to fully eliminate it.

Among the concerns are:

The sun and the wind energy are intermittent. The Sun doesn't always shine all the time – night and clouds -- and the wind doesn't always blow. The average power generated by a wind turbine is typically only about 1/3 of its peak capacity. That is, a 3 megawatt turbine only generates about 1 megawatt on average, and that's at a good location. A solar farm only generates about 1/4 of its peak capacity, on average, and that's at low latitudes. In northern regions, the average is considerably less.

The wind and solar generation often don't match grid electrical demand – either not supplying enough energy to the grid or generating energy that the grid doesn't need. Depending on weather conditions, season, location site, the inability to meet grid needs may occur daily for several hours, or even for periods of a week or longer.

Wind and solar generation is still a small fraction of total electric suppy, so the electric grid can make up for their intermittent supply nature using fossil fuel energy sources. If wind and solar are to replace fossil fuels, a low cost, efficient, practical method of storing large amounts of electric energy will have to be developed to match a time varying intermittent output from wind and solar to the time varying grid power demand. So far, the storage options are very expensive.

Wind turbine and solar farms require very large surface areas. For wind farms the average generated output power is on the order of 1.5 watts per square meter.(3) That is, 1 square Kilometer will generate only 1.5 megawatts. To generate 150,000 terawatt hours, the current world primary energy supply, would take over 4 million square miles of wind farms. To generate 250,000 terawatt hours, the world energy supply projected for 2040, would require 7 million square miles of wind farm areas – more than the land area of the US 48 continental states. There will be major environmental impacts. The land area for solar farms is less. Average generated electric output per unit solar farm area is on the order of 10 watts/m^2 about 6 times the output from windfarms. To supply 155,000 terawatt hours, the current world energy supply, would require a land area of 600,000 square miles, while to meet 2040 needs of 250,000 Terra Watt Hours, 1.5 million square miles.

Transmitting terawatts of energy over long distances form widely distributed wind and solar farms to increasingly urban populations would be very difficult and expensive. By mid-century, most of the world populations will be living in cities.

The amount of resources and construction required would be enormous to supply 155,000 terawatt hours of energy using wind turbines, for example, with an average output of 2 megawatts per turbine (1/3 peak capacity of 6 megawatts) would require construction of 8 million turbines. By 2040, 14 million turbines would be required. Implementing such large numbers of wind turbines in the time frame required to significantly reduce fossil fuel consumption would be very difficult. Similarly, building hundreds of thousands of square miles of solar power in the required time frame would be very difficult.

So, what new non-fossil-energy technologies can be developed that have the potential to supply clean, environmentally and economically practical energy for the world, enabling humanity to shift from fossil fuels? There are two technologies:

Beaming solar power form space satellites to receivers on Earth.

Generating power by Ocean Thermal Energy Conversion (OTEC) using the temperature difference between warm surface ocean water and deep cold water.

In this chapter, we describe how solar power satellites stations in Geosynchronous Orbit (GEO) could beam down thousands of megawatts to receivers located virtually anywhere on Earth, at much lower cost than generating power using fossil fuels or wind/solar renewable sources.

The OTEC power technology described in Chapter 6 is also very promising and can be used in combination with beamed solar power.

Low cost Maglev Launch to space technology, StarTram, is described in Chapter 3. In this chapter we summarize the space solar power and beaming technology that the StarTram launch system will enable. A more detailed description of the space power beaming system is available in StarTram: The New Race to Space, by Powell, Maise, and Pellegrino, and Silent Earth, by James Powell, Jesse Powell, and James Jordan.

What are the advantages of beaming solar power from space satellite power stations down to Earth, compared to generating in solar farms on Earth?

In space solar panels in geosynchronous orbit (GEO) see full sunlight 24 hours per day, compared to about 6 hours per day on Earth. As a result, 1 square meter of solar cells in space will generate 4 times as much energy as 1 square meter of solar cells on Earth.

Space power satellites are not subject to damage and degradation from storms and bad weather.

Space solar power stations in GEO orbit have continuous 24/7 view of receiver stations on Earth, to which they can continuously beam power at whatever demand level is required, which can vary throughout the day, week, month, or year.

Able to serve multiple areas around the planet, without the need for long transmission lines.

Much smaller land area on Earth required for beamed power receivers, compared to solar power systems on Earth.

Minimize amounts of energy required for human society, compared to other energy sources.`

FIGURE 4.1

ILLUSTRATES HOW POWER COULD BE BEAMED FROM THE SPACE SOLAR POWER SATELLITES TO EARTH.

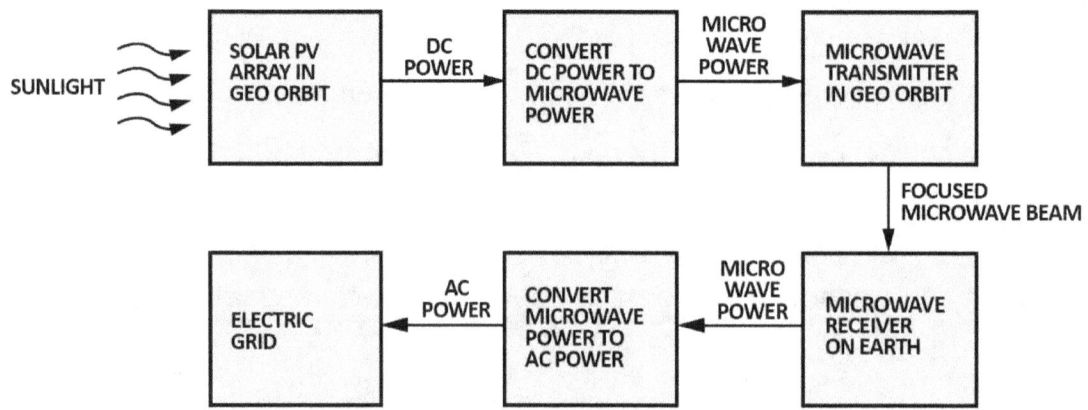

Laser pilot beams would be used to guide microwave beams to appropriate receivers on Earth, as illustrated in the following Figure 4.2

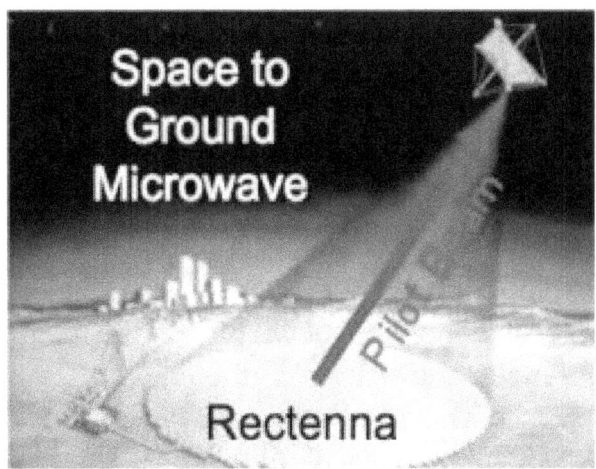

The various conversion processes, DC (Direct Current) to microwave power, microwave power beaming, and conversion of microwave power to AC (Alternating Current) power utilize well developed commercial technology: the efficiency of the conversion processes are typically: (5)

70% for DC to microwave power

93% for microwave beaming

85% microwave power to AC

Overall efficiency for DC power in GEO orbit to AC power on Earth will then be on the order of 0.70 x 0.93 x 0.85 =0.55. That is, for every 1,000 megawatt electrical of DC solar satellites power generated, the electric grids on Earth would receive 550-megawatts of AC (Alternating Current) power.

How big is the satellite PV array and its transmitter and the microwave receiver on Earth? The two questions are tied together by the physics of focusing microwaves. The diameters of the receiver and transmitter are related, plus the wavelength of the microwave beam and the distance between transmitter and receiver are related by the equation below.(6)

D_{TRANS} x D_{RCVR} = 2.44 λ_{BEAM} x $R_{Distance\ Trans\ to\ RCVR}$

For an SPS Satellite in GEO orbit, R = 3.58×10^7 meters

For a 10 gigahertz microwave beam

λ_{BEAM} = 3×10^{-2} meters

Incorporating R & λ, equation (1) is then

D_{TRANS} x D_{RCVR} = $2.44 \times 3 \times 10^{-2} \times 3.58 \times 10^7 = 2.62 \times 10^6 m^2$

Expressed in kilometers

DTRANS x DRCVR = 2.62 km²

If we make DTRANS the same size as DRCVR, then

DTRANS x DRCVR = 2.62 = 1.62 km or about 1 mile in diameter. If we DTRANS make bigger. E.g., with DTRANS = 2 kilometers, DRCVR is smaller= 1.3 kilometers

The sizes of the transmitter and receiver are independent of the amount of power that is transmitted, so it pays to transmit a lot of power per transmitter/receiver combinations. For an AC power of output of 2,000 megawatts(e) at a receiving station on Earth, one would have to generate 2,000/0.55 = 3,600 megawatts(e) of DC power in space, based on an end-to-end efficiency of 55%. The total area of the solar cell arrays to provide 3,600 megawatts(e) would be 9 km², based on 400 megawatts(e) per km² of area. Multiple PV arrays, each with a smaller area, could be connected to the transmitter.

To generate such large amounts of DC power, very large solar cell arrays will be required. Figures 4.3 and 4.4 show 2 solar power arrays, Sun Tower and Modular Sandwich, that have been studied by NASA.

FIGURE 4.3 SUN TOWER **FIGURE 4.4 MODULAR SANDWICH**

 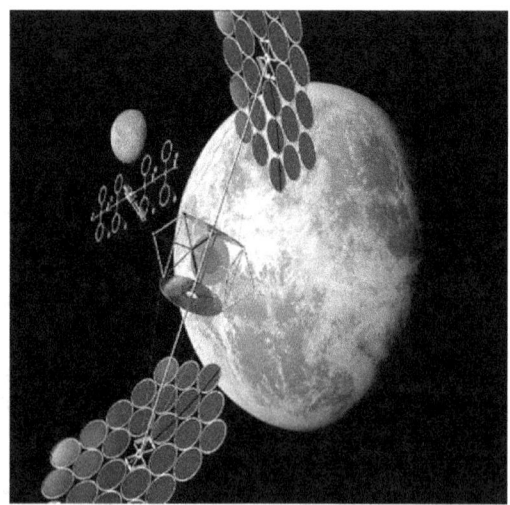

The solar PV array is just part of the total SPS mass in orbit. To assess total SPS mass and its cost, including launch cost, we use Mankins Design Reference Mission (DRM) #5 for the SPS-ALPHA concept described in his book, The Case for Space Solar Power (6). Mankins describes the SPS-ALPHA concept in detail – its various components, their mass and cost, how they are assembled in orbit, and so on. Five DRM versions of SPS-ALPHA are described, with increasing levels of beamed power to Earth. DRM#1 and DRM#2 are small scale demonstration SPS systems in LEO orbit. DRM #3 is a subscale integrated demonstration in GEO orbit. Two beamed power levels for DRM#3 are analyzed – one at 2 megawatts (e) and the other at 18 megawatts (e). DRM#4 is a commercial SPS system delivering 500 megawatts (e) to markets on Earth.

TABLE 4.1
PARAMETERS FOR 2000 MW SPS-ALPHA SPACE POWER SYSTEMS

Basis – DRM Case #5

Parameter	Value
Beamed Power Input Electrical Grid on Earth Per Unit	2,000 megawatts(e)
Location in Space	GEO orbit
SPS Platform Hardware Mass in Orbit per Unit	34,800 tonnes
DRM#5 Hardware Cost Per Unit	$5.7 Billion
Earth Receiver Cost Per Unit	$0.7 Billion
Total Hardware Cost (Launch Cost Not Included)	$6.4 Billion

DRM#5, the design we have selected, would deliver 2,000 megawatts(e) [2 gigawatts(e)] of power to markets on Earth. Table 4.1 summarizes the principal parameters for DRM#5

The total mass in GEO orbit of the 2,000 MW(e) DRM#5 Space Power System is 34,800 tonnes, corresponding to 17.4 Kilogram per Kilowatt of AC electric power delivered to grids on Earth. The total cost of its hardware in GEO orbit, not including launch cost, is 6.4 Billion dollars, corresponding to $3,200 per Kilowatt (e) amortized over 30 years, the corresponding cost per kilowatt hour is only 1.2 cents.

Beamed space solar power will be much cheaper than other generation options, if the solar power arrays can be launched into GEO at low cost.

Launch costs using rockets are much too great for space solar power to be economically practical. At current rates of approximately $10,000 per kilogram of payload to GEO, the launch cost for the DRM#5 System would be 350 Billion dollars, 50 times its space hardware cost. The 30 year amortized cost per kilowatt hour for launching DRM#5 would be 66 cents -- a completely impractical cost.

Using the StarTram launch system described in Chapter 3, the cost of placing payloads into GEO orbit will be much less than rocket launch. The amortized cost of the StarTram launch facility, including the cost of the spacecraft structures that holds the DRM#5 components will be on the order of $100 per kilogram of payload to GEO, a factor of 100 less than rocket launch. The corresponding launch cost amortized over 30 years of beamed space power would be 0.66 cents per kilowatt hour.

Adding together the cost of the solar satellite hardware and its launch cost, the total projected cost for electric power delivered to the grid is 1.2 plus 0.66 = 1.9 cents per kwhr(e) – only about 1/3 of the current average power cost of approximately 6 cents per kwhr(e).

The much lower cost of beamed space power will be very attractive to consumers, industrial, commercial, and residential. Its capability to reliably meet variable grid demand will be extremely attractive to utilities. Its capability to minimize long disruptive and visually objectional long distance transmission lines will be attractive to the public. The environmental cleanness of beamed space solar power – no pollution and greenhouse gas emissions, no objectionable power plants, no mining of oil, gas, and coal, no long coal trains and long oil and gas pipelines, will also be very attractive to the public.

The one issue the public will have questions about is the effect of the microwave beams on the environment. Is the microwave beam safe? Will it harm humans, animals, and birds that move through it? Based on studies of flora and fauna exposed to microwave, there appear to be no harmful effect. To assure safety the International Academy of Astronautics (IAA) has recommended in its 2008-2011 study of space solar power that the maximum intensity of wireless power transmission should be less than full summer sunlight at the equator, i.e., 1,000 watts/m2.(6) A 2 km diameter receiver receiving 2,350 megawatts of microwave power and delivering 2,350 x 0.85 = 2,000 megawatts of electric power to the grid, would have a microwave beam intensity at the receiver of:

$2,350 \times 10^6/(\pi/4) (2,000)^2 = 748$ watts/m^2, well under the IAA recommended limit. Microwave intensity outside of the receiver, which would be fenced off, would be much smaller. Typically, the microwave receivers would be located in low population density, non-arable regions. The land area required for the receivers will be less than 1/100th of the area required for solar farms on Earth.

How many Solar Power Satellites (SPS) will be needed in orbit to meet world energy needs – electric energy needs, transportation, heating, materials production, etc?

Starting with electrical generation, the world generated 20 Trillion KWH(e) of electrical energy in 2010. The EIA (The US Energy Information Agency) projects that the World Electrical generation will increase to 39 Trillion KWH (e) in 2040.(2) Extrapolating world electrical generation will further increase to approximately 45 Trillion KWH(e) in 2050. To supply all of the World's electrical power in 2050 would require 2400 SPS units in orbit, each of 2000 MW(e) capacity.

Even more SPS units will be required if synthetic gasoline, diesel fuel, and jet fuel, for cars, trains, and planes are manufactured using carbon dioxide from the atmosphere.

The economic benefits for solar power satellite (SPS) systems will be very substantial. The capital cost for the 2000 MW(e) DRM # SPS system, orbital hardware cost per launch cost, is projected to be 6.4+3.5 = 9.9 Billion dollars. The savings in power generation cost more than offset the SPS capital cost. The projected generation cost for SPS power is 2 cents per KWH(e) compared to the current average commercial generation cost of 6 cents per KWH(e).

The net cost savings per KWH(e) is then 6 cents/KWH(e) – 2 cents/KWH(e) = 4 cents/KWH(e). For 30-year operation of the 2,000 MW(e) SPS systems, total savings are 21 Billion dollars, 2 times greater than capital cost. The savings will be even greater as conventional energy costs grow in the coming decades.

Placing 2400 SPS units in orbit would require a total investment of 24 Trillion dollars. Spaced out over a period of 30 years, that's an annual investment of 0.8 Trillion dollars, a small fraction of annual world GDP (world Gross Domestic Product).

How much will beamed space power cost? Can it provide most of Earth's electric power? Based on Mankin's cost projections (6) for the SPS ALPHA-DRM#5 design and a StarTram launch cost of $100 per kilogram to GEO orbit, the parameters for the SPS units launched by StarTram are shown in the following Table 4.2.

TABLE 4.2
PARAMETERS FOR STARTRAM LAUNCH FACILITY TO PLACE 2,000 MW(E) SPS ALPHA-DRM# UNITS IN GEO

40 metric tonne payload per launch
12 launches per day
175,000 metric tonnes launched per year
5 SPS ALPHA – DRM#5 Units emplaced in GEO per year (2,000 MW(e) per unit, 34,800 tonnes per unit.
30 Billion dollars Capital cost of StarTram facility
$100/kg launch cost per kg to GEO (amortized)
3.5 Billion dollars cost to launch 1 SPS ALPHA-DRM #5 Unit [$1250/kw(e)]
First StarTram launches begin in 2025
2 potential rates of StarTram Facility Construction Studied:
Case 1: 2 facilities constructed per year in years 2025, 2026, 2027, 2028, and 2029 10 total Star Tram facilities constructed. Total Construction Cost of $300 Billion Dollars.
Case 2: 4 facilities constructed per year in years 2025, 2026, 2027, 2028, 2029. 20 total StarTram Facilities constructed. Total construction cost of $600 Billion dollars.
10 Billion dollars total capital cost for one 2000 MW(e) SPS Unit Space and Receiver Hardware Plus Launch Cost.
2 cents per KWH cost of SPS power beamed to Earth (Amortized Capital plus O&M costs)

For an annual world GWP of 200 Trillion dollars, and an annual amortized investment of 0.8 Trillion dollars the 2400 SPS units that would generate all of the estimated World's electrical power generation needs would cost less than 1 % of world GWP. Moreover, the annual savings of 1.8 trillion dollars in cheaper energy costs would more than offset the investment cost for the 2400 SPS units.

How soon can SPS systems be implemented? Implementation will require development of both the StarTram launch system and Solar Space Power array that can beam microwave power down to Earth.

We believe that with a very strong aggressive international effort that understands the critical need to achieve beamed solar power to save the environment, that SPS implementation could begin in 2025.

This seems, and is a very short time. However, looking back in history, radically new technologies have been developed in even shorter times. The US Manhattan Project began in 1942, with virtually zero knowledge of nuclear reactors and weapons. By 1944 we had high power reactors producing plutonium at Hanford, Washington, and centrifuges producing enriched uranium at Oak Ridge, Tennessee.

By July 1945, the US Trinity test had demonstrated the first atomic bomb. By September, 1945, we dropped nuclear bombs on Hiroshima and Nagasaki. In the next few years the US had hundreds of nuclear weapons and had developed the hydrogen bomb.

Another example, in 1962, President Kennedy initiated the Apollo Program to land a man on the moon. The 1st moon landing occurred 7 years later, in July 1969, after developing many radically new technologies including giant rockets, space capsules, and the moon lander including the telemetry so that Earth controllers could communicate with and control the system.

Assuming that SPS launches could begin in 2025, what would be the schedule for its implementation? Table 4.2 gives the parameters for a StarTram launch facility to start placing 2,000 megawatt SPS units in GEO.

The capital cost to put 2400 SPS units in orbit to supply all of the World's power in 2050 AD, is 24.6 Trillion dollars including the cost of constructing 20 StarTram launch facilities. The Net World Savings from SPS power at 2 cents per KWH, as compared to purchasing conventional power at 6 cents per KWH is 20.6 Trillion dollars over the period from 2025 to 2050AD. The savings in the cost of power is comparable to the cost of putting the SPS units in orbit. And, as noted above, it is likely that SPS units will become even cheaper as the technology evolves, further increasing the savings.

World GDP is presently 65 Trillion dollars annually and probably will increase substantially in the coming decades. Projections vary as to what the increases will be for the various nations of the World. Using the predictions of Price Waterhouse Coopers (7) for the 20 largest economies in the World, Table 4.3 lists the top 10 ranked countries in 2014 and 2050AD.

China's GDP grew by a factor of 5 from 10.0 in 2014 to 50.9 in 2050. India grew by a factor of 10.5 from 2.0 in 2014 to 21.0 in 2050. The 2 most populous countries on Earth, with approximately 1/3 of the total World population in 2050. European countries – France, Germany, and United Kingdom – grew only by a factor of about 2. Italy dropped out of the top 10, also growing only by a factor of 2. Japan grew by only a factor of 1.6 and the U.S. grew by a little more than a factor of 2. Russian and Brazil grew by a factor of 3.

TABLE 4.3

	2014 Annual GDP			2050 Annual GDP	
	Country	GDP (in 2014 Trillions USD)		Country	GDP (in 2014 Trillions USD)
1.	U.S.	17.5	1.	China	50.9
2.	China	10.0	2.	U.S.	38.0
3.	Japan	4.8	3.	India	21.0
4.	Germany	3.9	4.	Japan	7.8
5.	France	2.9	5.	Brazil	7.4
6.	United Kingdom	2.8	6.	Russia	6.2
7.	Brazil	2.2	7.	Germany	6.2
8.	Italy	2.2	8.	United Kingdom	5.9
9.	Russia	2.1	9.	Mexico	5.8
10.	India	2.0	10.	France	5.7

China, the US, and India will be the dominant economic powers in 2050, with their combined GDPs being almost 60 percent of total annual World GDP, which will be on the order of 200 Trillion dollars, 3 times greater than today's (2014) annual World GDP.

And many other developing nations in Africa, Asia, and South America will want to catch up to the big GDP nations, so 2050 is not a final state of affairs. Given low cost, environmentally clean SPS power, they will seek to use it to develop their economies and a higher standard of living, so that electrical power demand will still continue to substantially increase over the remaining 2050 to 2100 period of the 21st Century.

Each StarTram launch would place 40 tonnes of payload into GEO orbit, one of the many modules that would be joined together in orbit to form the DRM#5 solar array and microwave transmitter that would deliver 2,000 megawatts of electric power to receiving fields then to grids on Earth.

On StarTram launch facility would have the capability to launch 5 DRM#5 solar power systems into orbit annually. 20 StarTram launch facilities could put 2,000 DRM#5 units into orbit over a 20 year period.

Based on the 2 cases, case 1, 10 total StarTram launch facilities, and case 2, 20 total StarTram launch facilities, the implementation schedule for worldwide beamed solar power can be projected, starting in 2025. Following the completion of Phase 1 development.

Based on a 2025 start date for SPS launches, implementation schedules have been analyzed (Table 4.4) for Phase 2.

Case 1: 10 Total StarTram launch facilities constructed in years 2025 through 2029, 2 per year.

Case 2: 20 total StarTram launch facilities constructed in years 2025 through 2029, 4 per year.

As each StarTram launch facility is completed and begins operation, each year it launches 5 SPS units into GEO orbit. Table 4.4 gives the total SPS units in orbit at the end of year 2029. For Case 1, 10 total StarTram launch facilities, there are 150 SPS units in orbit. For Case 2, 20 total StarTram launch facilities, there are 300 SPS units in orbit.

Already in 2029, the SPS units supply a significant fraction of the total World electric power generation of 32.1 Trillion KWH. For Case 1, SPS supplies 2.7 Trillion KWH (Table 4.5), 8.4 percent of the World Total. For Case 2, SPS supplies 5.40 Trillion KWH (16.8 percent).

Going forward from 2030 in Phase 3 the already constructed StarTram launch facilities continue to launch more SPS units, increasing the amount of SPS power beamed to Earth.(Table 4.4). In 2050 AD, Case 1 has 1260 SPS units in orbit, supply 48% of total World electric generation while Case 2 has 2400 SPS units in orbit supplying 96% of total World generation.

The corresponding annual investments for the SPS units, together with the annual savings in World electric power cost, based on 2 cents /KWH average cost compared to 6 cents per KWH for power from conventional sources -- coal, gas, nuclear, wind, and ground solar PV – are shown in Table 4.5.

By 2035 the net annual savings offset the net annual investment in SPS units. From then on the net annual savings are greater than the annual investment. By 2050, for Case 1 the total SPS investment over the period 2025 is 12.30 Trillion dollars, while the total savings in power cost is 10.14 Trillion dollars, resulting in a net cost for the period of 12.30-10.14 = 2.16 Trillion dollars. By 2050, for Case 2, total investment is 24.60 Trillion Dollars, total savings are 20.6 Trillion dollars, and net total costs are 4 Trillion dollars.

Total annual GDP in 2050 for the top 20 largest economies in the World is projected to be 188 Trillion dollars, expressed in 2014 US dollars (7). China's annual GDP is 50.8 Trillion USD, America is 38.0 Trillion USD, India, 20.8 Trillion USD, accounting for more than ½ of the 190 Trillion US dollars total. The other 17 economies, Japan, Russia, Brazil, Germany, etc., down to #20, Switzerland at 2.0 Trillion dollars annually, make up the balance. It is noteworthy that China will have a bigger GDP than America, and that India will have a DDP that is more than 1/2 that of the US.

In the years following 2050 Case 2 will have a net annual savings in power cost of about 45 Trillion KWH x $0.04/KWH=1.8 Trillion dollars, about 1% of World Annual GDP. Not included in the savings enabled by the SPS-StarTram programs are the much greater external costs of generating electric power from fossil fuels – the costs of global warming from greenhouse gas emissions, strip mining, oil spills, pollutants that damage health, contamination of aquifers from fracking for gas and oil, and so on. The benefits from eliminating the external costs are even greater than the economic benefits.

TABLE 4.4
IMPLEMENTATION SCHEDULE FOR SPS ALPHA DRM#5 UNITS IN ORBIT AND SPS POWER BEAMED TO EARTH.

Basis: 5 SPS Launches Per Year from 1 Star Tram Facility
2,000 MW(e) per SPS Unit

EIA Projections for World Power Demand ( )

Year	Number of StarTram Facilities		SPS Launches per year		Number of SPS Units In Orbit		SPS Power Supply Trillion KWH/year		Total World Power Demand
	Case 1	Case 2	Case 1	Case 2	Case 1	Case 2	Case 1	Case 2	Trillion KWH/YR
2025	2	4	10	20	10	20	0.18	0.36	29.5
2026	4	8	20	40	30	60	0.54	1.08	30.1
2027	6	12	30	60	60	120	1.08	2.16	30.8
2028	8	16	40	80	100	200	1.80	3.60	31.4
2029	10	20	50	100	150	300	2.70	5.40	32.1
2030	10	20	50	100	200	400	3.60	7.20	32.7
2035	10	20	50	100	450	900	8.1	16.2	35.8
2040	10	20	50	100	700	1400	12.6	25.2	39.0
2045	10	20	50	100	950	1900	17.1	34.2	42.0
2050	10	20	50	100	1200	2400	21.6	43.2	45.0

Case 1: 10 StarTram Launch Facilities – SPS supplies 48% of Total World Power Demand

Case 2: 20 StarTram Launch Facilities – SPS Supplies 96% of Total World Power Demand

TABLE 4.5

INVESTMENT IN AND SAVINGS IN COST OF BEAMED WORLD POWER FROM SPS ALPHA – DRM# SOLAR SATELLITES IN GEO ORBIT

Year/ Period	Number of SPS Units in Orbit		SPS Investment During Year/Period In Trillion Dollars			SPS Power Savings During Year/Period in Trillion Dollars	
	Case 1	Case 2	Case 1	Case 2		Case 1	Case 2
2025	10	20	0.16	0.32		0.005	0.01
2026	30	60	0.26	0.52		0.02	0.04
2027	60	120	0.36	0.72		0.045	0.18
2028	90	180	0.46	0.92		0.07	0.28
2029	150	300	0.56	1.12		0.11	0.44
2030-2034	400	800	2.50	5.00		1.08	4.32
2035-2039	650	1350	2.50	5.00		1.98	7.92
2040-2044	900	1800	2.50	5.00		2.88	11.52
2045-2050	1200	2400	3.00	6.00		3.95	9.90
Total SPS Investment in Trillions $			12.30	24.60	**Total SPS Power Savings in Trillions $**	10.14	20.6

Net Savings from SPS Power Savings – Total SPS Investment

Case 1: Net Savings = 12.30 T$ - 10.14 T$ = -2.16 Trillion $

Case 2: Net Savings = 24.60 T$ - 20.60T$ = -4.00 Trillion $

Summarizing, Case 2 enables the World to meet virtually all of its electrical pwer needs by 2050, with the total integrated savings in the cost of delivered electrical power almost equal to the capital cost for implementing beamed space solar power. After 2050, the savings in cost of electrical power per KWH(e) will be much more than the capital cost of the beamed space power.

Chapter 4

List of References

1. World Energy Consumption, https://en.wikipedia.org/wiki/ world_energy_consumption

2. World Energy Scenarios – composing World Energy Scenarios to 2050—World Energy Council
https://www.worldenergy.org/publications/2013/world-energy-scenarios-composing-energy-futures-to-2050/

3. National Wind Watch Size of Industrial Wind Turbines https://www.wind-watch.org/faq-size.php

4. Topaz Solar Farm, https:en.wikipedia.org/wiki/Topaz_Solar_Farm

5. Space Based Solar Power, en.wikipedia.org/wiki/space_solar_power

6. The Case for Solar Power, John C. Mankins, Virginia Edition Publishing, LLC (2014)

7. List of IMF Ranked Countries by Past and Projected GDP (nominal)
https://en.wikipedia.org/wiki/List_of_countries_by_past_and_projected_GDP_(nominal)

List of Figure Credits

Figure 4.1 Illustrates How Power Could Be Beamed from the Space Solar Power Satellites to Earth, author, Powell

Figure 4.2 Space to Ground Microwave Usiig Laser Pilot Http:/en.wikipedia.org/wiki/file: space_to_ground_microwave_pilot_beam.png, Author, NASA

Figure 4.3 Sun Tower, http://en/wikipedia.org/wiki/file:Suntower, Author, NASA

Figure 4.4 Molecular Sandwich,

http://en-wikipedia.org/wiki/File:Solar_Power_satellite_sandwich_or_abbawa_concept.jpg,Author, NASA

Chapter 5

Stopping Runaway Warming of Spaceship Earth

First, the facts. The concentration of carbon dioxide in the world's atmosphere is steadily rising. Figure 5.1 (1) shows a graph of concentration in the atmosphere, measured at Mauna Loa in Hawaii, rising from 315 ppm in 1960 to 400 ppm in 2016.

FIGURE 5.1

CO_2 CONCENTRATION IN THE ATMOSPHERE, MEASURED AT MAUNA LOA IN HAWAII

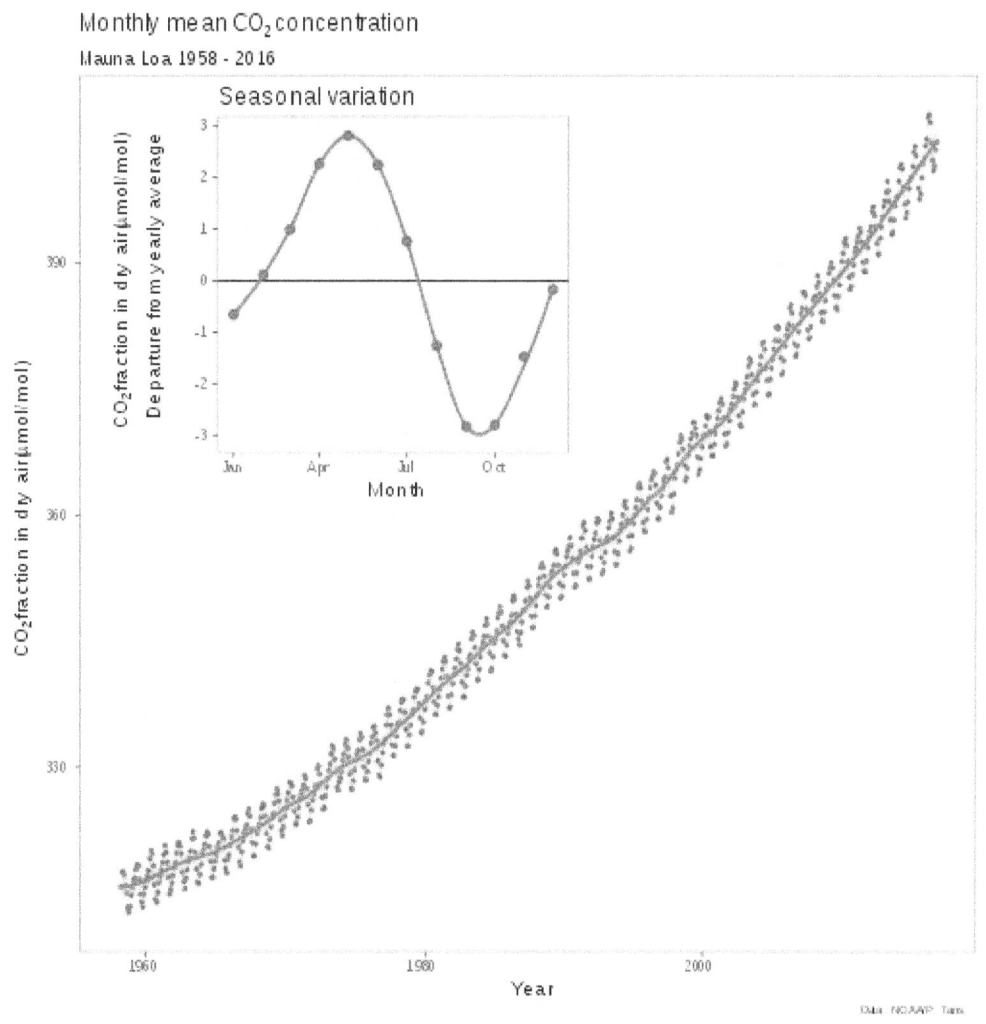

The rise in atmospheric carbon dioxide concentration began in 1800, as shown in Figure 5.2 (2), when the world started massive consumption of fossil fuels at the beginning of the Industrial Revolution. The data from measurements of air inside ice cores at Law Dome in Antarctica show that before 1800, CO_2 concentration was virtually constant at 280 ppm.

FIGURE 5.2

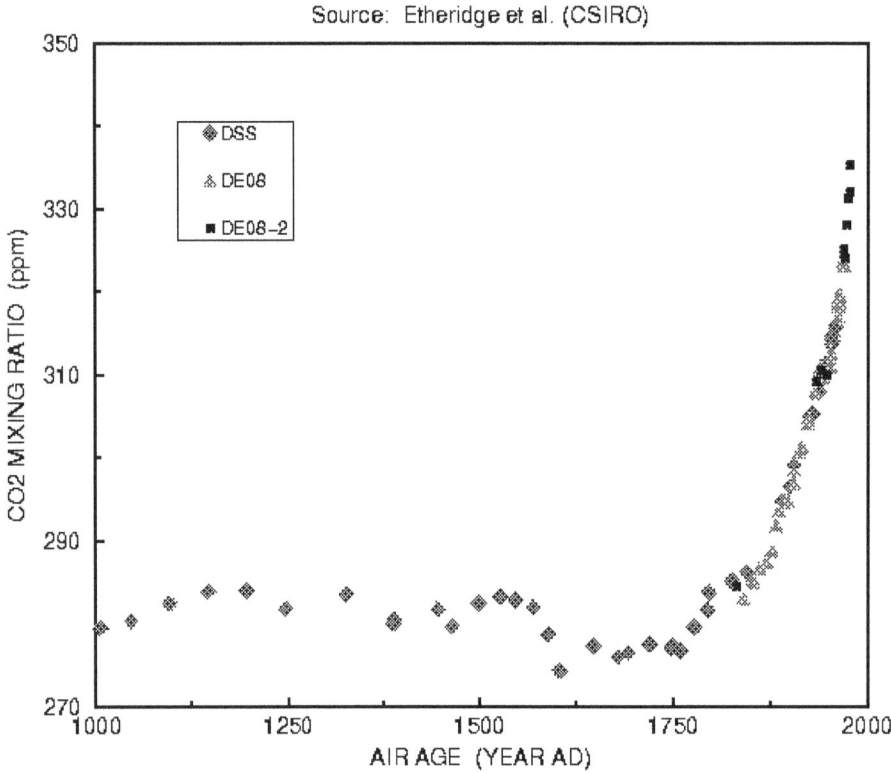

Other measurements of the atmospheric CO_2 concentration taken from Antarctic ice cores show in Figure 5.3 (3) that the CO_2 level has not risen over 280 ppm in the last 650,000 years. The CO_2 concentration did vary from a low of 180 ppm to a high of 280 ppm, but over a long period, 5,000 years. Today, in our human world, we have gone up by 100 ppm in CO_2 concentration in only 60 years.

Over the same period of rise from atmospheric CO_2 concentration from the pre-industrial level of 280 ppm to the present 400 ppm, World carbon dioxide emissions from burning fossil fuels have increased in step with atmospheric concentration.

This is hardly surprising. If you dump a pollutant into a lake, its concentration in the lake goes up. Before humans started burning fossil fuels, the CO_2 from decomposing plants and animals went into the atmosphere, to be extracted from it to make new plant and animals. When massive amounts from the combustion of fossil fuels goes into the atmosphere new plants and animals can't handle it, and the CO_2 concentration in the atmosphere steadily increases.

FIGURE 5.3 HISTORICAL EVIDENCE OF CO_2 IN EARTH'S ATMOSPHERE

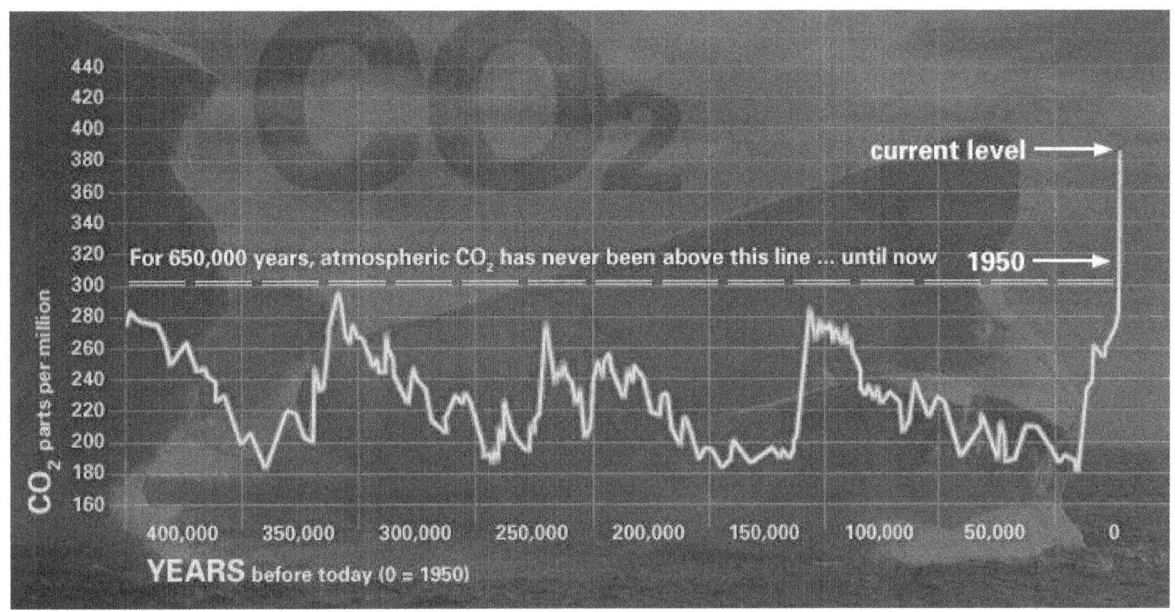

Figure 5.4 (4) shows global carbon dioxide emissions from 1850 to 2030. In 1950, just 65 years ago global CO_2 emissions were only 5,000 million metric tonnes (5 Billion tonnes) of CO_2 annually. Today, in 2016, they are 35 Billion tonnes per year. 7 times greater, and projected to rise to 43 Billion tonnes annually by 2030.

FIGURE 5.4
GLOBAL CARBON DIOXIDE EMISSIONS FROM 1850 TO 2030

https://www.c2es.org/docUploads/global- CO2-emissions-historical.png

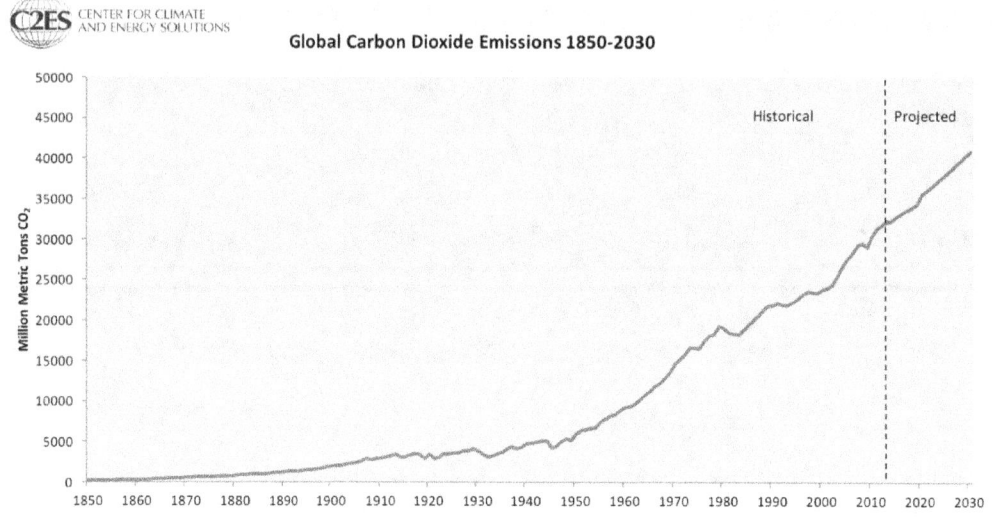

The rise in CO_2 concentration in future years will probably be much greater than projected as the developing countries like China and India rapidly industrialize and consume more fossil fuels. Figure 5.5 (5) shows the CO_2 emissions from China, from 1980 to 2008. In the very brief period from 2002 to 2008, CO_2 emissions jumped from 3,000 million metric tonnes annually to 7700 million metric tonnes, a factor of 2.5 increase.

China and India constitute almost half of the World population. Adding in the other developing countries on the order of 1/2 of World population will seek to substantially grow, which will require much greater energy use.

FIGURE 5.5
CARBON DIOXIDE EMISSIONS FROM CHINA

https://upload.wikimedia.org/wikipedia/commons/0/0e/Carbon_dioxide_emissions_due_to_consumption_in_China.png

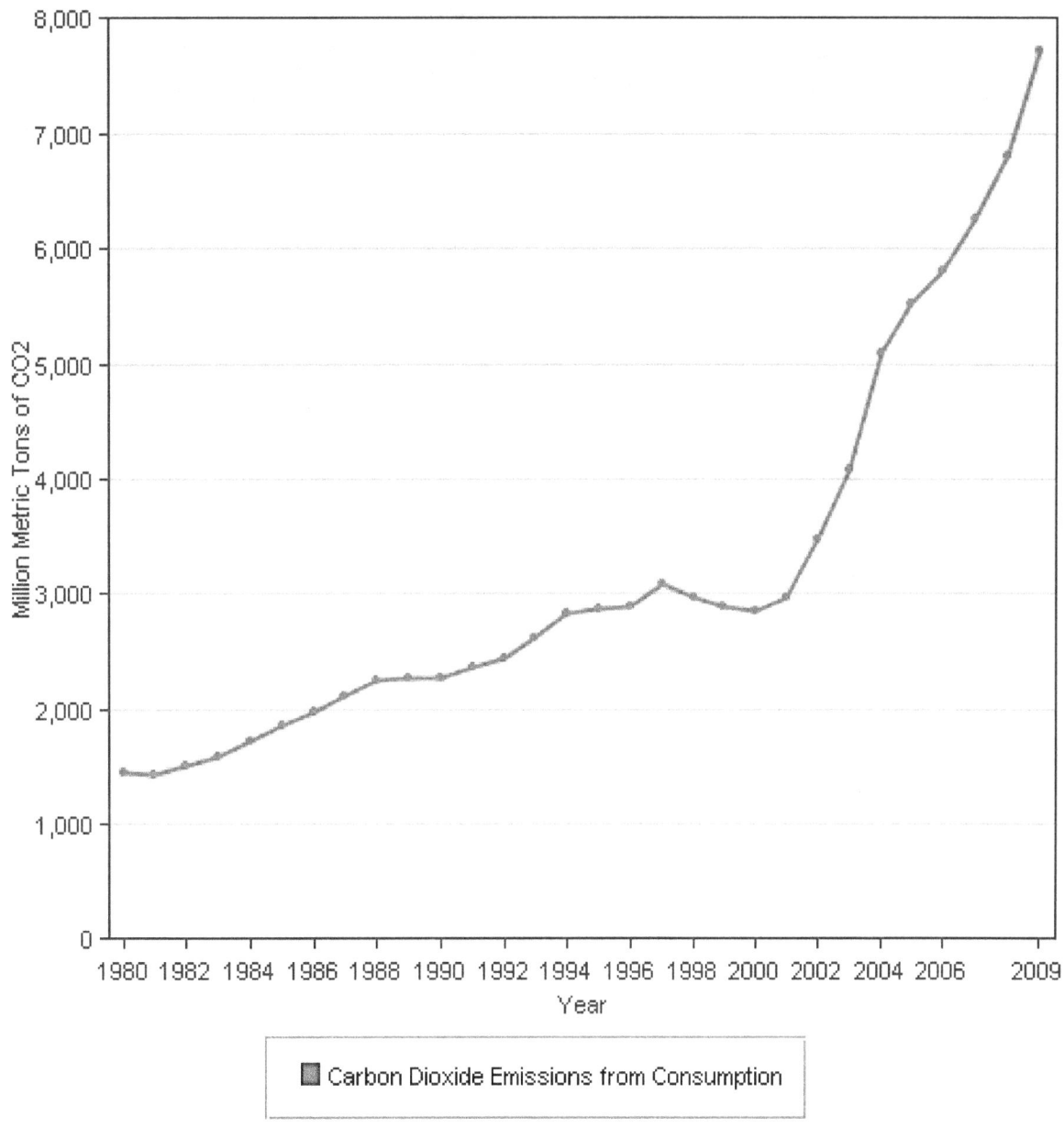

If the World continues to rely on fossil fuels for the bulk of its energy needs in the coming decades, what will the result be in atmospheric CO_2 concentration and global temperature?

The recent 5th Assessment Report of the IPCC (International Panel on Climate Change) (6) has forecast future atmospheric CO_2 concentrations and global temperatures based on various scenarios involving "business as usual" using fossil fuels and different potential schedules for shifting from fossil fuels to clean, renewable energy sources.

Figure 5.6 (7) shows projected CO_2 atmospheric concentrations from 2000 to 2100, for four emission pathways. The top pathway assumes global CO_2 emissions will continue to rise to 2100. The bottom pathway assumes global CO_2 emissions will be done after 2020.

The top pathway forecasts a CO_2 concentration of 1400 ppm by 2100, while the bottom projects 500 ppm, a factor of almost 3 difference. Which pathway will humanity take?

The corresponding global temperatures increases depend strongly on the CO_2 concentration for the top pathway, global temperature would increase by 4 degrees centigrade (7.2 degrees Fahrenheit relative to the temperature of 2000 AD. For the bottom pathway, the temperature increase would only be about 1 degree Centigrade (1.8 degree Fahrenheit).

Most climate scientists believe that human society could survive a global temperature rise of 2 degrees centigrade (3.6 degrees Fahrenheit) although there will be very harmful consequences – droughts and wild fires, rising sea levels, sever storms, and coastal flooding, major crop failures, increased deaths from heat waves, etc.

FIGURE 5.6

Projected Atmospheric Greenhouse Gas Concentrations

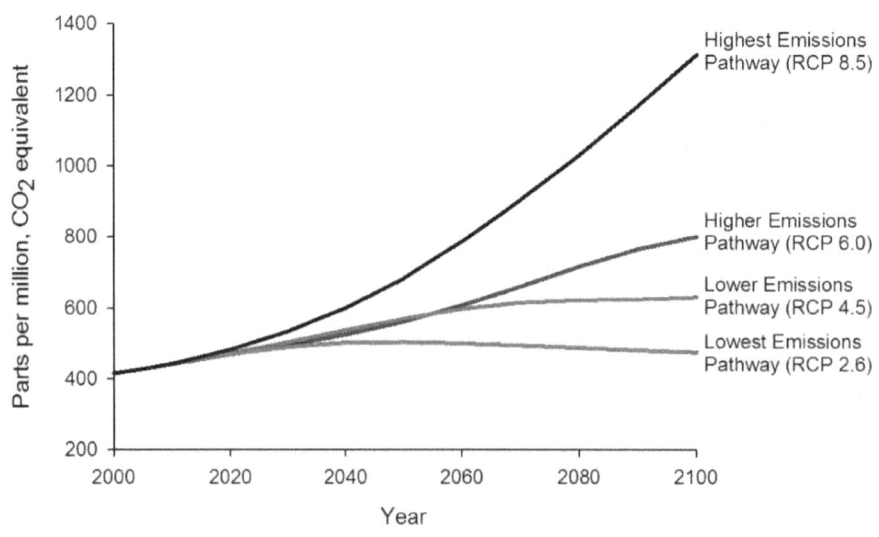

It is very sobering to look back 50 million years ago to the Eocene era, when the global CO_2 concentration was 800 ppm. There were pine forests in the Antarctic, crocodiles in the Arctic, and sea levels were 300 feet higher than today.

Given all the facts, only a tiny fraction of which are mentioned here, its hard to understand why many persons deny:

a. That the climate is changing
b. If they accept that the climate is changing, they deny it is due to humanity and fossil fuels.

Well, the facts are:

a. Global CO_2 concentration is increasing
b. Increasing global CO_2 concentration correlates with the CO_2 emitted by burning fossil fuel
c. That CO_2 in the atmosphere can increase global temperature

Point c) is proven both by historical data over millions of years, where temperature and CO_2 concentrations track each other, and by standard industrial practice.

The temperature of a thermally radiating body in a furnace depends on the CO_2 concentration in the gases inside the furnace. Increasing the CO_2 concentration in the surrounding gas increases the temperature of the radiating body.

Why? Because infrared radiation from the radiating body gets absorbed in the CO_2 in the gas, with about half radiated back from the CO_2 to the radiating body. The temperature of the body then has to go up to get rid of the radiant energy that the CO_2 deposits on it.

Sophomore chemical engineering students learn how to calculate the temperature of a radiating body in a furnace as a function of CO_2 concentration in the gas by going to the proper chart and equations using Perry's Chemical Engineering Handbook.

It is very clear that humanity must transition from fossil fuels to clean, sustainable, renewable energy sources, and to make the transition as soon as possible, if environmental catastrophe and the crash of Spaceship Earth are to be avoided.

How well is the transition proceeding? Not very well. Countries make promises and sign statements setting future goals, but the actual transition is proceeding slowly. Moreover, the present renewable energy options have economic and operational problems tghat hinder their implementation. These problems are discussed in the Appendix.

Beaming power down to Earth from solar power satellites in space, as described in Chapter 4, is a very promising new clean sustainable energy option, with the potential for worldwide, unlimited, continuous electdric energy at substantially lower cost than present electric generation methods.

In parallel with implementing beamed power from space, implementing the capability to extract CO_2 from the atmosphere and burying it in geologically stable formations would greatly help to avoid catastrophic global warming.

Extracting CO_2 from the atmosphere is quite possible, and has been demonstrated in many studies. The real challenge is to do it in extremely large amounts at acceptable cost.

To significantly impact CO_2 concentration in the atmosphere, a CO_2 extraction rate should be comparable to the current CO_2 emission rate, i.e. on the order of 30 Billion tonnes per year.

To be economically practical, a goal of $100 per tonne of extracted CO_2 would have an annual cost of 3 Trillion dollars. A tremendous amount, but still much less than the current world GDP of approximately 80 Trillion dollars, and the 250 Trillion dollar world GDP projected for 2050 AD.

3 Trillion dollars per year for CO_2 removal compared to $250 Trillion per year for World GDP in 2050, if it prevents global environmental collapse, would appear to an acceptable choice.

The CO_2 extraction system described in the Appendix has a total annual cost of approximately 1 Trillion dollars per year for 30 Billion tonnes of CO_2 extraction annually.

Various technology approaches have been proposed for CO_2 extraction from the atmosphere.[8]

- Bio-energy with carbon capture & storage
- Biochar
- Enhanced weathering
- Ocean Fertilization
- Direct air capture

The first technology, bioenergy, is not practical for the scale needed to significantly reduce the CO_2 concentration in the atmosphere, i.e. 30 Billion tonnes of CO_2 per year. The decay of World land vegetation releases about 450 Billion tonnes of CO_2 per year, 15 times greater.

To achieve a negative carbon dioxide emission of 30 Billion tonnes annually using the bioenergy approach would require harvesting 6% of the existing World land vegetation every year transporting it to power plants, drying it, burning it, and extracting CO_2 from the flue gas.

The world's total mass of vegetation is estimated to be 2.4 Trillion metric tonnes expressed as dry weight. In terms of natural weight as harvested, with water content, total mass is on the order of twice as great, about 5 trillion tons.

Harvesting and transporting 10 percent of 5 Trillion tons per year, 300 Billion metric tonnes, of vegetation to capture 30 Billion tonnes of CO_2 is not practical. The costs would be enormous, with the ton-miles required much greater than the World's present world transport capability.

The second technology, biochar, uses a fraction of the potential combustion energy of biomass to pyrolyze the biomass at high temperature, producing charcoal, which is then buried in topsoil. The amounts of biomass to be handled to sequester the equivalent of 30 billion tonnes of CO_2 would be hundreds of Billions of tons per year, comparable to Bio Energy – far too great to be practical.

The third technology, *enhanced weathering*, by itself does not remove carbon dioxide from the atmosphere. Rather, it is a method of sequestering CO_2 in geologic stable rock formations, one CO_2 has been removed from the atmosphere. The Carb Fix project in Iceland(8) injected 250 tonnes of CO_2 into underground basalt formations solidifying it into calcite in 3 years. The CO_2 was injected as pressurized carbonated water at 25 tonnes of water per tonne of CO_2.

The fourth technology, ocean fertilization proposed putting various nutrients – iron, phosphorous, nitrogen to stimulate the growth of phytoplankton and capture CO_2. However, the efficiency of the added nutrients I producing more phytoplankton is low. Moreover, when the phytoplankton die and oxidize, their carbon is released back into the ocean.

Finally, ocean fertilization may have very dangerous and destructive effects on the oceans very complex and diverse ecological systems, with the risk of destroying many marine species and greatly altering ocean life.

The best approach appears to be direct capture of CO_2 from the atmosphere with injection of the extracted CO_2 into underground rock formations for geologic storage. Its environment effects would be minimal compare to alternative approaches, and in fact, would have tremendous environmental benefits if it could help keep the atmospheric CO_2 concentrations at a safe level.

A wide range of direct capture methods has been studied and experimented with. Two overall approaches are being investigated.

Reaction of CO_2 in air with solid or liquid chemicals to form compounds that contain the CO_2. For example, calcium oxide will absorb CO_2 from air that is mixed with steam to form calcium carbonate (limestone). The calcium oxide is regenerated by heating the limestone to 1000 degrees centigrade.

Absorption of CO_2 on solid polymeric resin material with subsequent desorption of the CO_2 using changes in humidity or temperature.

Approach #1 is being tested by Carbon Engineering at a pilot plant in Squamish, British Columbia, Canada.(9) Using a solution of potassium hydroxide in water that flows inside a cooling tower where it extracts atmospheric air. The CO_2 in air reacts with the potassium hydroxide to form potassium carbonate. Calcum hydroxide is then added to the solution, forming solid calcium carbonate and regenerating the potassium hydroxide. The solution is then processed to recover from solid carbonate pellets that are then heated to high temperature to drive off carbon dioxide, to be later sequestered in underground rock formations.

The calcium carbonate is heated to 900°C by burning natural gas to recover the carbon dioxide extracted from the atmosphere. Burning the natural gas emits more CO_2 so that fossil fuels are still required, which emits more CO_2. This, and the complex engineering and the expensive process equipment required, are major disadvantages for approach #1.

Approach #2 contacts atmospheric air with a polymeric resin material which absorbs CO_2 from the air. When the resin reaches the CO_2 saturation point, the CO_2 can be desorbed, either by heating the resin, or humidifying it, depending on the type of resin used. The desorbed resin can then absorb more CO_2, functioning in a absorption/desorption cycle.

Experiments carried out by ZChang and colleagues on absorption of CO_2 on PEI resin particles are described in their papers, "Capturing CO_2 from Air Using a Polyethylene – Silica Absorbent In Fluidized Beds",(10) and in "Performance of Polyethaneimine – Silica Absorbent for Post Combustion CO_2 Capture In a Bubbling Fluidized Bed", Chemical Engineering Journal (11)

The experiments demonstrated that PEI particles could absorb CO_2 from atmospheric air at levels up to 5% by weight in repeated cycles, with absorption at ambient temperatures and desorption at 80 degrees centigrade. Absorption was demonstrated both in static settled beds and bubbling fluidized beds of small PEI particles.

The diameter of the PEI particles was 250 microns, with 60% of their weight being amorphous substrate and 40% being polyethylene deposited on the substrate.

A wide variety of absorbents have been studied, including silica particles with many kinds of resins, nanofillibrated cellulose, etc. For this study the most promising material appears to be PEI particles studied by Zhang, et al, in their experiments on absorption of CO_2 from the atmosphere.(10)

The Zhang et al study focused on a design that used a bubbling fluidized bed of PEI particles to absorb CO_2 from the atmosphere. The bubbling fluidized bed approach requires complex equipment, a large fluidized bed coupled to a large heated desorbing chamber, coupled to a particle recirculatory and airflow system. An issue is the potential damage and degradation to the PEI particles as they circulate through the system experiencing impacts and collisions.

A simpler, less complex and expensive approach for absorption of atmospheric CO_2 is to use multiple static beds of PEI particles through which air flows at ambient temperature and pressure. When the PEI particles reach their saturation point, airflow ceases. The beds are then closed off and heated to desorb pure pressurized CO_2 which is collected and removed for disposal in geologic rock formation.

This approach is described in detail in the Appendix, for a CO_2 extraction rate from the atmosphere of 30 Billion tonnes annually, comparable to the present emission rate of CO_2 into the atmosphere from the combustion of fossil fuels.

The CO_2 extraction system described here appears practical in terms of its technology and economics. Total capital cost is projected to be 9.5 Trillion dollars. Amortized over 30 years, the annual amortization cost would be 320 Billion dollars. The annual operating cost – personnel, energy, and maintenance – is projected to be 540 Billion dollars per year, for a total $29 per tonne of CO_2 extracted.

The 860 Billion dollar annual cost is a very large expenditure, but acceptable in terms of other World expenditures. Current, World military expenditures are 1.6 Trillion dollars per year. World GDP is on the order of 80 Trillion dollars per year, about 100 times greater. World GDP is projected to grow to over 200 Billion dollars annually in 2050, more than twice the CO_2 removal cost.

The economic cost of removing 30 Billion tonnes of CO_2 appears acceptable, given that it will help to prevent environmental collapse from global warming.

Chapter 5

List of References

1. https://en.wikipedia.org/wiki/global_warming
2. www.CSIRO.au/greenhouse.gases/
3. Climate change vital signs of the Planet: Evidence

http.//climate.nasa.gov/evidence

4. Global Carbon Dioxide Emissions 1858 – 2030

www.C2es.org/facts.figures/international - emissions/historical

5. Carbon Doxide Emissions From China

http://en.wikipedia.org/wiki/List_of_countries_by carbon_dioxide_emissions

6. Projected Atmospheric Greenhouse Emissions

http://www.epa.gov/climate-change-science/

7. Observed and Projected Changes in Global Average Temperature

https://www.epa.gov/climate-change-science

8. https://en.wikipedia.org/wiki/carbon_dioxide_removal

9. Productivity of the Worlds Main Ecosystems LE Rodin, MI Bazilevich, NN R0ZOV

https://books.google.com/books?id=jzkrAAAAyAAJ

10. MIT Technology Review, Go Inside an Industrial Plant That Sucks Carbon Dioxide Straight Out of the Air, Peter Fairly, June 6, 2016

11. Capturing CO2 from ambient air using a polyethyleneimine–silica adsorbent in fluidized beds Wenbin Zhang, Hao Liun et al, Chemical Engineering Science, 116, 6 September (2014) pages 306-316

12. Performance of Polyethyleneimine—Silica Adsorbent for Post-Combustion Capture n a Bubbling Fluidized Fluidized Bed, Chemical Engineering Journal Zhang, et al, 251 1 Septermber 2014, pages 293-303.

List of Figure Credits

Figure 5.1 Mauna Loa CO_2 Monthly Mean Concentration, https://commons.wikimedia.org/wiki/File:Mauna_Loa_CO2_monthly_mean_concentration.svg, Author Delorme

Figure 5.2 Atmospheric Carbon Dioxide from Law Dome Ice Cores, Source, CSIRO

Figure 5.3 Historical Evidence of CO2 in Earth's Atmosphere,

https://commons, Wikipedia.org/wiki/files: Evidence_CO_2.jpg, Author NASA

Figure 5.4 Global Carbon Dioxide Emissions, 850-2030, https://www.c2es.org/facts-figures/international-emissions/historical

Figure 5.5 Carbon Dioxide Emissions From China

https://en.wikipedia.org/wiki/List_of countries_by carbon_dioxide_emissions

Figure 5.6 Projected Atmospheric Greenhouse Emissions

https://www.epa.gov/climate-change-science

Figure 5.7 Observed and Projected changes in Global Average Temperatures, https://www.epa.gov/climate-change-science

Chapter 6

Food and Water for Spaceship Earth

Degradation and loss of the capability to grow adequate food supplies due to drought and loss of soil and soil fertility has been a major cause of the collapse of numerous societies throughout human history. In many cases, the failed societies responsibility for their inability to grow sufficient food, by employing agricultural methods that erode topsoil, degrade soil fertility, and deprive soil of adequate water and nutrients.

Historical examples of agriculturally failed societies include the classic Mayan civilization in the area now known as Mexico, due to slash and burn agriculture (1), 2300 BC Mesopotamia located between the Tigris and Euphrates rivers, due to salting of its topsoil by irrigation water from the rivers(2), Easter Island, the Anasazi in the American Southwest, and Norse Greenland as described in Jared Diamond's book, Collapse (3).

The Easter Island population grew too big to be supported by the limits on agricultural capacity, the Anasazi cut down trees for buildings, resulting in reduced water supply from the mesas, and the Norse Greenlanders cut down forests for wood fires, cut up turf for buildings and over grazed the land.

Today, the World appears to be on the same path of agricultural collapse that has caused previous societies to fail. Mario-Helena Sevedo of the UN Food and Agricultural Organization (FAO) told a forum on World Soil Day that about 1/3 of the World's soil has already been degraded, and if current rates of degradation continue all of the World's topsoil could be gone within 60 years with the global amount and productive land per person in 2050 will be only a quarter of the level in 1960.(4)

If we are to keep up with global food demand, the UN estimates that the world needs 148 million acres of new farmland every year.(5) However we currently losing 30 million acres of farmland every year because of soil degradation.(6)

Soil degrades from multiple causes:

- Erosion from storms, and flooding, type irrigation
- Increased salinity forms minerals in irrigation water from rivers and underground aquifers
- Depletion of soil nutrients due to over use
- Contamination from fertilizers and insecticides
- Compaction of soil by overpumping of underground aquifers.
- Additional stresses that affect food supply include
- Droughts
- Lower crop yields due to higher temperature resulting from global warming

Increasing global temperatures will have major impacts on crop yields. Figure 6.1 shows the effect of 1°C, 2°C, and 4°C increases in global temperature from the baseline of 2000 AD for 5 crops (7), due to the effects of higher temperature on reproductive capability.

- Africa maize
- Asia rice
- India wheat
- US maize
- US soybean

With a 2°C rise in global temperature, crop yield will decrease about 20%, with a 4°C rise, crop yield will decrease by 40%.

FIGURE 6.1

PROJECTED CHANGES IN YIELDS OF SELECTED CROPS WITH GLOBAL WARMING

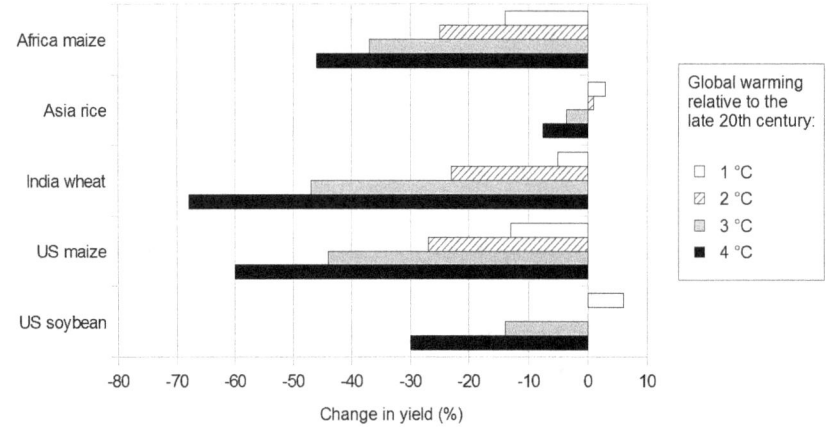

Increasingly, droughts due to global warming are threatening the world food supplies. South Africa has the worst drought in 20 years, dramatically reducing the corn and sugar production for its 54 million inhabitants, as well as the water levels behind its dams.(8)

North Korea has its worst drought in a century. 80% of the rice seedlings in the South and North Hwangbae province have dried up and water reservoirs are very low.(8) Parts of Brazil have the worst drought in 50 years. San Paulo inhabitants do only have 40% their normal water supply, with tap water only available for 2 hours per day.(8) In the Caribbean, Puerto Rico and the Dominican Republic are experiencing severe water shortages, both for crops and personal use.

Many countries are overpumping underground aquifers to irrigate their farmland because of droughts. According to NASA(9), twenty one of the world's largest aquifers, located in China, india, the US, France, and other countries, are pumping more water from them, than is replaced by rainfall.

Overpumping is not sustainable. The water table keeps dropping, forcing wells to be drilled deeper and deeper to reach water. In many area, water tables have dropped to depths of 300 feet or more, making it very difficult to pump out water.

Moreover, as the water tables drop, so does the level of the ground surface. Parts of the Central Valley surface in California have dropped 30 feet due to overpumping.

In addition, as soil compacts due to overpumping aquifers, it becomes difficult for rainfall to penetrate through the soil to refill the aquifer.

World-wide, underground aquifers supply 35% of the water used by humans, California is presently getting 60% of its water from underground aquifers.

Besides overpumping aquifers and putting future crops at risk of even more severe lack of water when the aquifer gives out, irrigating with water from underground aquifers increases the salinity of soil from the dissolved minerals in the water as it evaportates. Increasing soil salinity reduces crop yields.

California and the Western US is a prime example of how drought affects farms. Figure 6.2 shows a map of the drought areas in California in October, 2014. 92% of the State was in the D-3 and D-4 conditions, extreme drought and exceptional drought.

FIGURE 6.2
MAP OF CALIFORNIA DROUGHT STATUS

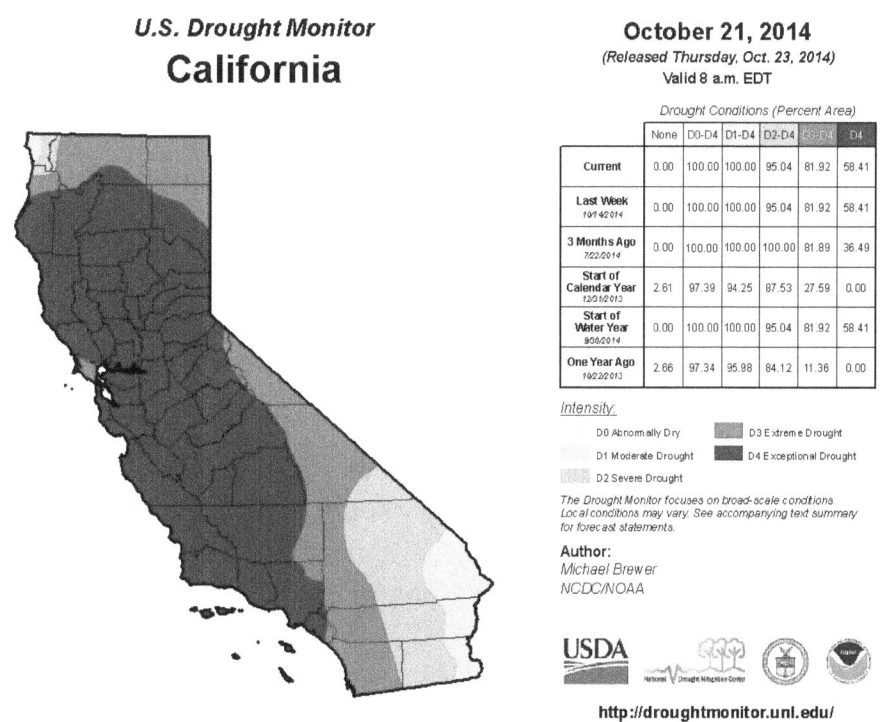

Figure 6.3 shows a map of the drought areas in the US in August 2015. In addition to California, extreme and exceptional drought areas are found in parts of Nevada, Utah, Oregon, Washington, Idaho, and Montana.

FIGURE 6.3
DROUGHT AREAS IN THE US IN AUGUST 2015

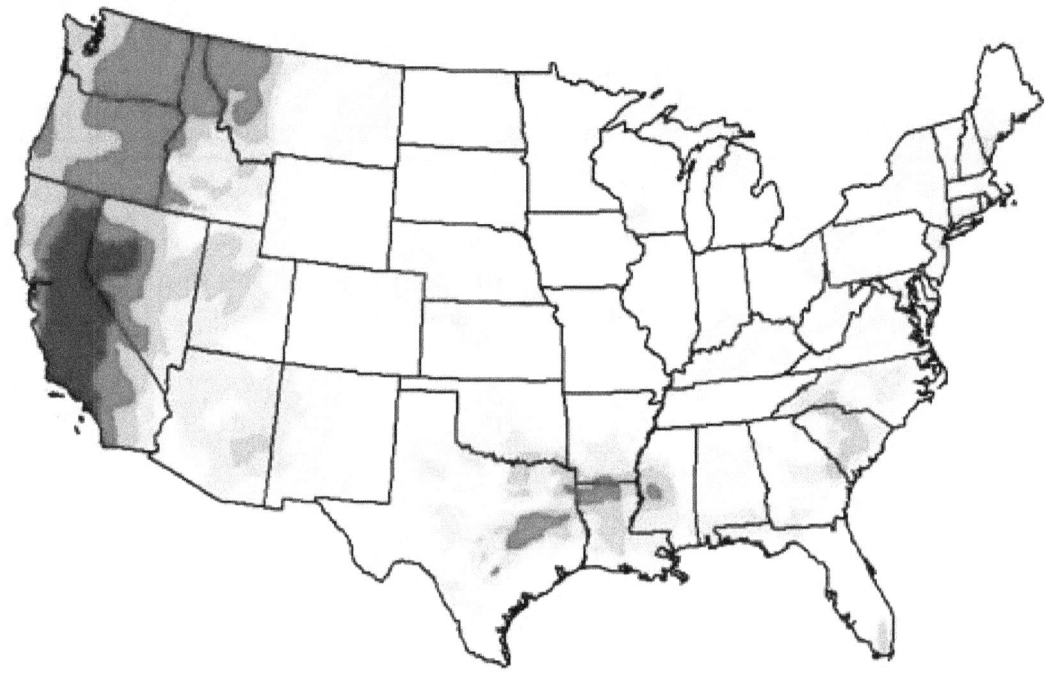

What can be done to prevent soil loss and degradation and provide the water needed to ensure sustainable food supplies for an increasing world population projected to grow to 9 Billion persons by 2050 and 10 Billion by 2100 AD, with increased droughts due to global warming?

Key to this goal is the development of new technologies that can produce massive quantities of low cost freshwater that can be transported to drought area for farming and residential applications.

Natural water supplies are too limited to meet this goal. California, Nevada, Arizona, Colorado New Mexico, Utah, and Wyoming already use virtually all of the rivers flow. Since the late 1990s the Colorado River has only made it once to its discharge point at the Gulf of California – the rest of the time the final river bed has been dry.

And the river's flow in decreasing. Since 1998, the water volume in Lake Mead behind the Hoover Dam has decreased 60%. The situation is similar all over the World, with natural sources of water insufficient to mitigate the effects of increasing droughts.

A very promising new technology for low cost fresh water is to desalinate sea water using Ocean Thermal Energy Conversion (OTEC), in which the thermal energy in warm surface water at 25 to 30 degrees Centigrade is used to evaporate sea water producing fresh water which is condensed from vapor to liquid by heat exchange with cold water at 5 degrees centigrade drawn from deep ocean water layers thousands of feet below the ocean surface.

Desalination plants already operate around the world using various processes to remove the salt from sea water, including multiple effect distillation, reverse osmosis , and electrodialysis.

Currently, total world desalination is 87 million cubic meters per day of fresh water (10) At 264 gallons per cubic meter, that's equal to 23 Billions of gallons per day. The typical cost for desalination is about $2 to $4 per 1,000 gallons (10). At an average cost of $3 per 1,000 gallons, worldwide desalination costs about $70 million dollars daily, equal to 25 Billion dollars, annually.

The California water system manages 49 cubic kilometers of water annually, 51% for environmental uses, 39% for agricultural use, and 11% for urban use (11). California has a population of about 39 million, corresponding to a daily per capita total water usage of 3.4 cubic meters (900 gallons) per day. That's a lot of water. At $3 per 1,000 gallons and 11% urban usage, it would be affordable for residents.

However, 3$ per 1,000 gallons is not affordable for farmers. For most crops at 30 cents per 1,000 gallons, the cost of irrigation water equals the value of the crops it would produce.(12) To meet the need for water for crops, the cost of desalinated sea water must be substantially less than 30 cents per 1,000 gallons.

Table 6.1 shows the cost of the water needed to grow various crops as a function of the cost to produce the need water, for production costs of 30 cents per 1,000 gallons, 1 dollar per 1,000 gallons, and 8 dollars per 1,000 gallons.(12)

TABLE 6.1
AMOUNT AND COST OF WATER TO GROW SELECTED FOOD CROPS

Basis: The World's Water, Vol. 7, Peter Gleick, Pacific Institute (2009) [3].

Food Crop	Average Gallons of Water to Produce Unit Amount of Selected Food Crop*	Cost of Water For Crop as Function of Cost of Water for 1000 Gallons		
		$0.30	$1.00	$8.00
Milk (1 qt)	1000	$0.30	$1.00	$8.00
Hamburger (1 lb)	1800	$0.54	$1.80	$14.40
Chicken (1 lb)	520	$0.16	$0.52	$4.16
Egg (1)	400	$0.12	$0.40	$3.20
Potato (1 lb)	120	$0.04	$0.12	$0.96
Wheat (1 lb)	165	$0.05	$0.165	$1.32
Rice (1 lb)	480	$0.15	$0.48	$3.84
Apple (1 lb)	80	$0.02	$0.08	$0.64

*Note: Data for water required for the above crops is taken from The Worlds Water, Vol. 7, where it is given in liters per kilogram, and converted to gallons per pound. If a range of water consumption values is given, the average value is calculated for the above table.

At $3 per 1,000 gallons, the water for 1 quart of milk would cost $3.00, an unaffordable amount, 1 pound of hamburger would cost $5.40, also not affordable. 1 egg would cost $1.20 per egg, $14.40 for 1 dozen eggs, just for water.

The basic process for desalinating seawater using ocean thermal energy conversion is shown in Figure 6.4. Warm surface water at approximately 25 degrees centigrade flows into an evaporator, where a small fraction of the water, about 1 percent, evaporates at low pressure. The remaining 99% of the surface water input, at a few degrees lower in temperature, is discharged back into the ocean.

FIGURE 6.4

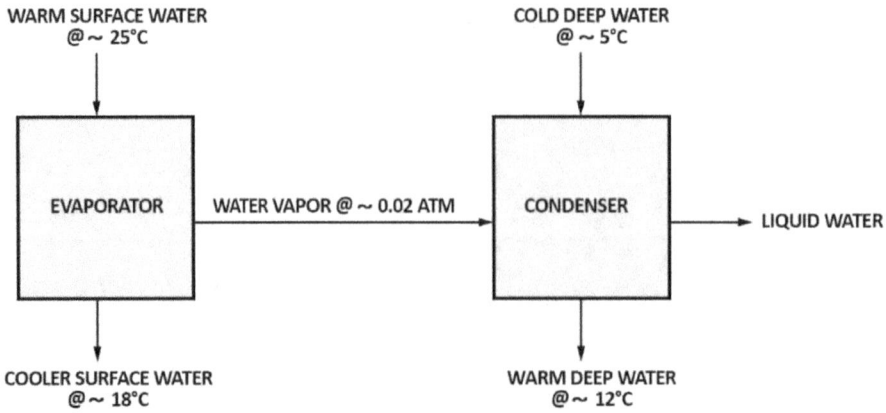

DESALINATION OF SEA WATER USING OCEAN THERMAL ENERGY

The water vapor at a low pressure of approximately 0.02 atmospheres, is condensed to liquid water by heat exchange with cold deep water at 5 degrees centigrade drawn from the deep cold layer thousands of feet below the ocean surface. The latent heat of condensation heats the deep cold water by a few degrees centigrade, with the warmed deep water discharged back into the ocean.

The OTEC desalination process does not require energy for the actual desalination, but does require energy for pumping the warm surface water and deep cold water flows, with flow rates that are on the order of 100 times greater than the liquid fresh water output.

Using ocean thermal energy to produce electric power and/or fresh water is a very old idea.

In Jules Verne's, Twenty Thousand Leagues Under the Sea, published in 1870, Captain Nemo proposed generating electricity from the difference in temperature difference between the ocean's warm surface and the deep cold water a kilometer beneath it, using a thermoelectric device.

Eleven years later, the French inventor, D'Arsonval, stimulated by Jules Verne, proposed the closed OTEC power generation cycle. Warm surface water would vaporize ammonia, which would expand through a turbine to generate electricity. The ammonia vapor would then condense on heat exchangers cooled by deep cold water.

Subsequently, D'Arsonval's student, George Claude, proposed the open OTEC power cycle. Instead of vaporizing ammonia, water vapor from the warm surface water expands through the turbine generator. The exhaust water vapor from the turbine is then condensed using cold deep ocean water.

Over the decades since D'Argonval and Claude, there has been continuing interest in OTEC. Small pilot plants have demonstrated net power production. The largest plant tested to date generated a gross power output of 255 kilowatts electric with a net power output of 103 kilowatts after deducting power for operating equipment (pumps, etc.). The open cycle OTEC plant operated for 6 years (1993 to 1998) in Hawaii.(13)

Previous work on OTEC has primarily focused on electric power generation, though the possibility of desalination using open cycle OTEC has also been studied.

Though OTEC is attractive for power and desalination, the existing proposed OTEC systems are far too expensive to be implemented. For example, Table 6.2 shows parameters for a representative 50 Megawatt closed cycle OTEC plantship. (14) The OTEC plantship displacement is 120,000 tonnes, over twice the 52,000 tonnes displacement of RMS Titanic. Using conventional ship technology, 50 MW(e) about the maximum power that could be generated by an OTEC plantship.

To generate 1000 MW(e), the standard output from conventional power plans, one would need 20 OTEC plantships. To generate todays world electric generation of 20 Trillion KWH(e) annually, one would need 46,000 ships, an impossibly high number.

TABLE 6.2
PARAMETERS FOR 50 MW(E) CLOSED CYCLE OTEC PLANTSHIP

Source: First generation 50 MW(e) OTEC Plantship for the production of electricity and desalinated water, L. Vega and D. Michaelis (2)

- Ammonia working fluid
- 198 meter plantship length
- 39 meter plantship beam
- 16 meter draught
- 24 meter depth
- 120,600 tonnes displacement
- 5 modules of 16 MW(e) gross capacity
- 5 seawater/ammonia evaporator modules, each 7100 m3 volume (34 m long, 13 m wide, 16 m high)
- 5 ammonia turbo-generator units, each 580 m3 volume (12 m long, 8 m wide, 5 m high)
- 5 seawater/ammonia condenser modules, each 7100 m3 volume (34 m long, 13 m wide, 16 m high)
- 73,900 m3 total volume of 5 modules on 50 MW(e) plantship
- Dimensions of RMS Titanic
    - 269 meter length
    - 28 meter beam
    - 10.5 meter draught
    - 19.7 meter depth
    - 52,000 tonnes displacement
- 246 m3/sec warm water flow rate
- 138.6 m3/sec cold water flow rate

The closed cycle OTEC Plantships for electric power generation will also be very expensive. For a deadweight tonnage of 120,000 tonnes, the capital cost of the ship alone will be well over 100 million dollars, on the order of $3,000 per KW(e), just for the ship.

The cost of open cycle OTEC plantships for electric and/or desalination based on convention ships will be even greater because of the tremendous volume required for flow of the low pressure, 0.02 atm, water vapor.

The total volume of the power generation system modules on the 50 MW(e) plantship shown in Table 6.2 is 73,900 cubic meters, about 60% of its 120,000 tonne displacement. The 50 MW(e) open cycle OTEC plantship design (14) has a much larger displacement, 247,000 tonnes, twice as big as the 50 MW(e) closed cycle OTEC plantship.

A new approach for OTEC power and freshwater generation is proposed, based on thermally insulated large ice structures that float in the ocean. The cost of manufacturing and insulating these large ice structures is very low, even for ice island structures that are a kilometer or more in diameter.

The recent book, The Ice World Cometh, by James Powell, Jesse Powell, and John Powell, available at Amazon.com, describes in detail how large ice structures can be constructed and thermally insulated, their projected capital and operating costs, refrigeration power requirements, and their potential applications, both ocean and land based.

Attractive applications of large ice structures, in addition to OTEC-ICE electric power generation and desalination, include low cost, high coastal barriers to protect against rising sea levels, storm surges, and tsunamis, low cost off-shore housing, airports, seaports and wildlife refuges, freezing toxic waste dumps to prevent pollution, protective structures to enclose hazardous industrial sites, including nuclear, water reservoirs in drought and wildfire areas, transport of fresh water over long distances in low cost ice pipe structures and many other uses.

For OTEC-ICE electric and desalination applications, Figures 6.5 shows a diagram of a typical large ice island structure in the ocean. The insulation and refrigeration parameters for a 1 square kilometer area ice island are given in Table 6.3. The insulation and refrigeration costs for the 1 square kilometer ice island are kept low by using low cost insulation materials, e.g. bagged mixtures of sand and perlite, and CO_2 gas mattress layer on the island's bottom surface, and deep cold water for refrigeration and heat sink of the - 30° C/+5° C refrigeration cycle, which provides an attractive coefficient of performance, 3.49 watts(th) of refrigeration capacity per watt of electric power to the refrigeration cycle.

The capital cost of freezing the ice to manufacture large ice islands is also very low, on the order of $2 per cubic meter. For the 1 square kilometer, 60 meter thick ice island, total volume is on the order of 60 million cubic meters, or about 120 million dollars.

The very large volume and low cost ice island structures enables OTEC-ICE plantships to generate electric power and fresh water at very low cost.

FIGURE 6.5
TYPICAL LARGE ISLAND STRUCTURE.

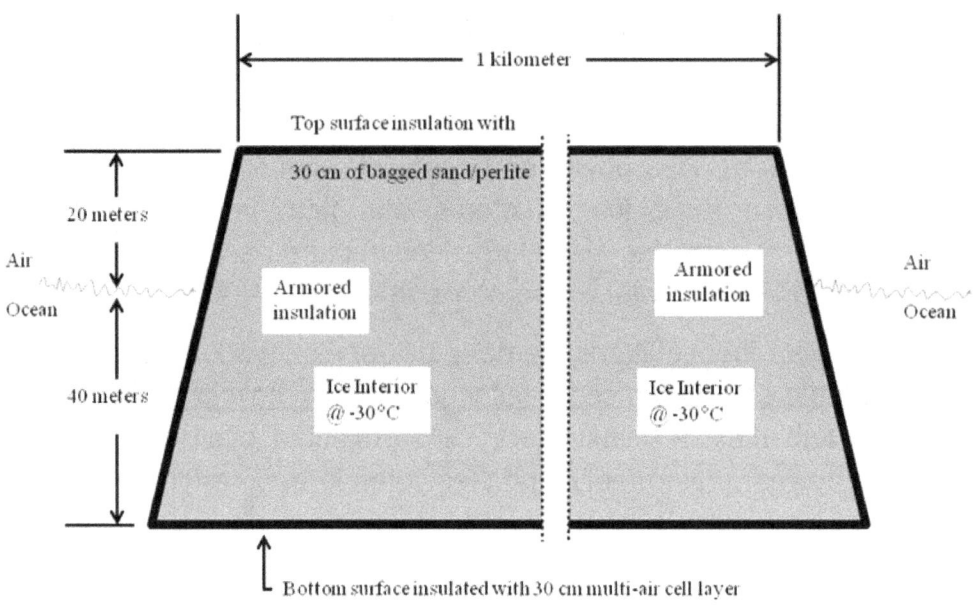

TABLE 6.3
INSULATION AND REFRIGERATION COSTS FOR LARGE ICE ISLAND

Basis: One km2 surface area, 60 meters thick
$\Delta T = 35°C$ across insulation, COP = 3.49 watts(th)/watt(e)

Surface			
Parameter	Top	Bottom	Side
Surface area	10^6 m^2	10^6 m^2	0.24×10^6 m^2
Type and thickness of insulation	Dry sand/perlite, 30 cm	1 CO_2 air mattress, 30 cm	Armored insulation
Thermal input thru insulation/m2	9.9 watts(th)/m^2	1.8 watt(th)/m^2	16.3 watts(th)/m^2
Refrigeration power/m2	0.284 watts(e)/m^2	0.52 watts(e)/m2	4.67 watts(e)/m^2
Total refrigeration power	2.84 MW(e)	0.52 MW(e)	1.12 MW(e)
Total annual refrigeration cost for surface	1.24 M$/year	0.22 M$/year	0.48 M$/year
Total annual refrigeration cost for all surface	1.94 M$/year		
Cost of insulation/m2	$15.5	$15	$153
Total cost of insulation for surface	15.5 M$	15 M$	37 M$
Amortized annual cost of insulation (30 years) for surface	0.52 M$/year	0.50 M$/year	1.23 M$/year
Annual amortized cost for all surface	2.25 M$/year		
Total annual ref cost and amortized cost for all surface	4.19 M$/year		

Figure 6.6 illustrates the layout of a 2,000-megawatt(e) OTEC-ICE plantship using forty 50 MW (e) closed cycle ammonia modules of the type show in Table 6.2. Parameters for the 2,000 MW(e) OTEC-ICE plantship are given in Table 6.4.

FIGURE 6.6
ARRANGEMENT OF 50 MW(E) UNITS ON 2000 MW(E) OTEC ICE PLANTSHIP.

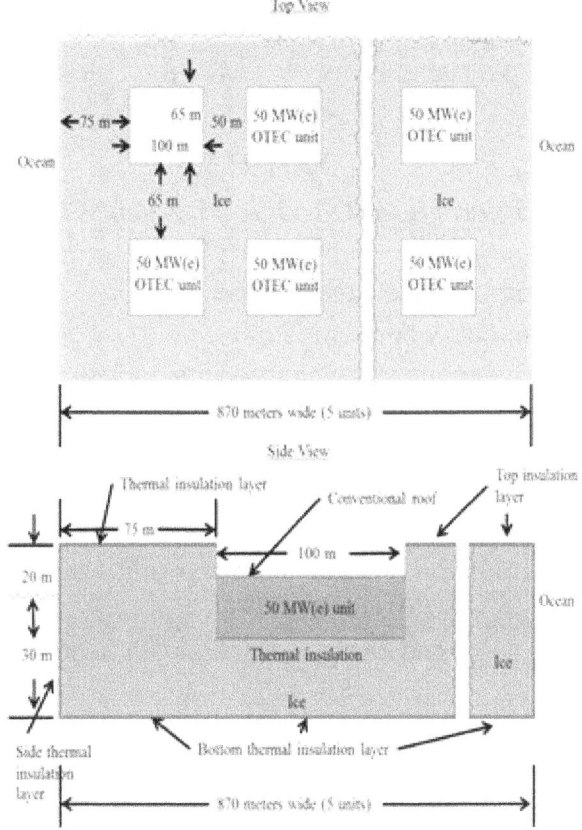

TABLE 6.4   DESIGN PARAMETERS FOR 2000 MW(E) CLOSED CYCLE OTEC-ICE PLANTSHIP.

- 2000 MW(e) net output
- 3200 MW(e) gross output
- Net/gross output ratio = 0.62
- Dimensions of 50 MW(e) net output unit
- 104 m long, 65 m wide, 16 m high
- 6800 m^2 floor area of 50 MW(e) unit
- 40 units (50 MW(e) each) on OTEC-ICE plantship
- 270,000 m^2 total floor area for 40 units
- Dimensions of OTEC-ICE plantship
- 1150 m long, 870 m wide, 50 m deep
- 1×10^6 m^2 bottom surface area of OTEC-ICE plantship

- 10,500 m³/second warm water input at 26° C
- 5500 m³/second cold water input at +5° C
- 50/50% sand/perlite insulation for top surface
- CO_2 gas cell insulation for bottom surface
- Armored dry sand insulation for side surface

Net electric output is 2000 MW(e) after deducting power for pumps, refrigeration, and other equipment from the 3200 MW(e) gross electric output. Table 6.5 gives the refrigeration power and cost for the OTEC-ICE plantship. Total refrigeration power for the to, side, and bottom surfaces of the 1150 meter long, 870 meter wide, and 50 meter deep OTEC-ICE plantship is only 5.4 MW(e), a small fraction of the 3200 MW(e) gross electric output.

TABLE 6.5
REFRIGERATION POWER AND COST FOR -30° C ICE INTERIOR OF 2000 MW(E) OTEC-ICE PLANTSHIP

Basis: +26° C warm surface water temperature
+5° C deep cold water temperature
Deep cold water used as intermediate coolant between 26° C and -30° C

-30° C ice interior insulation has outer temperature of +5° C and inside temperature of -30° C; $\Delta T = 35° C$
3.49 COP

Parameter	Plantship Top Surface Under 50 MW(e) Units	Plantship Top Surface and Unit Walls	Plantship Bottom Surface	Plantship Side Surface
Insulation type	Dry sand bags	Sand/perlite bags	CO_2 gas cells	Armored dry sand
Area, m2	270,000	944,000	1,000,000	202,000
Thickness, cm	30	30	30	30
k, w/mK	0.14	0.08	0.0155	0.14
g(th), w(th)/m2	16.9	9.9	1.8	16.9
P(e), w(e)/m2	4.67	2.84	0.52	4.67
Total power, MW(e)	1.26 MW(e)	2.66 MW(e)	0.52 MW(e)	0.93 MW(e)
Annual cost $M/year (5 cents/KWH)	0.54 M$/year	1.66 M$/year	0.22 M$/year	0.40 M$/year
Total power for all surfaces = 5.37 MW(e)				
Total refrigeration power cost = 2.32 million $/year				

Capital cost parameters for the 2000 MW(e) OTEC-ICE plantship are summarized in Table 6.6 The Capital costs for freezing and thermal insulation total on the order of 140 million dollars. Adding in the 150 million dollars for the pipes that bring deep cold water to the OTEC-ICE plantship, the total capital cost of the plantship is 290 million dollars. Amortized over 30 years, plantship capital cost corresponds to only 0.055 cents per KWH(e) of output.

The much bigger capital cost component is for the ammonia closed cycle equipment, at $5,000 per KW(e). Amortized over 30 years, this corresponds to 1.9 cents per KWH(e) of output. Adding the amortized capital cost of the OTEC-ICE plantship and the power cycle equipment and the refrigeration power cost, total power cost is only 2 cents per KWH(e) for 2,000 MW(e) output.

TABLE 6.6
CAPITAL COST OF 2000 MW(E) OTEC-ICE PLANTSHIP

Basis: Included are capital costs of energy and equipment for freezing ice structure, insulation layers on surface of ice structure and refrigeration equipment, and ice pipes to draw up cold water.

The capital cost of the ammonia (NH_3) closed cycle power equipment, heat exchangers, water pumps and anchor cables are assumed to be $5000/KW(e).

Parameter	Top Surface	Bottom Surface	Side Surface
Insulation cost/m2	15.46 $/m^2	15 $/m^2	153 $/m^2
Area m2	1.21 x 10^6	1 x 10^6	202,000
Capital cost (million dollars)	19	15	30
Total capital cost for insulation = 65 million $			
Total ice volume = 37 million m^3			
Capital cost for freezing (@ $2/m^3) = 74 million dollars			
Total length of cold water ice pipes = 12 kilometers (8 pipes, 22 m ID, 1.4 km long pipe)			
Total capital cost of cold water pipe @ 12.5 M$/km = 150 million $			
Total capital cost of 2000 MW(e) OTEC-ICE plantship = 290 million $			
Equivalent cost per KW(e) = 145 $/KW(e)			
30 year amortized cost/KWH = 0.055 cents/KWH			
30 year amortized cost/KWH for $5000/KW(e) OTEC cycle equipment = 1.9 cents/KWH			

Turning to using OTEC-ICE plantships to desalinate seawater, Figure 6.7 shows the flowsheet for a plantship that produces 2 Billion gallons of fresh water daily.

FIGURE 6.7
FLOW SHEET FOR OTEC-ICE PLANTSHIP TO PRODUCE 2 BILLION OF FRESH WATER PER DAY.

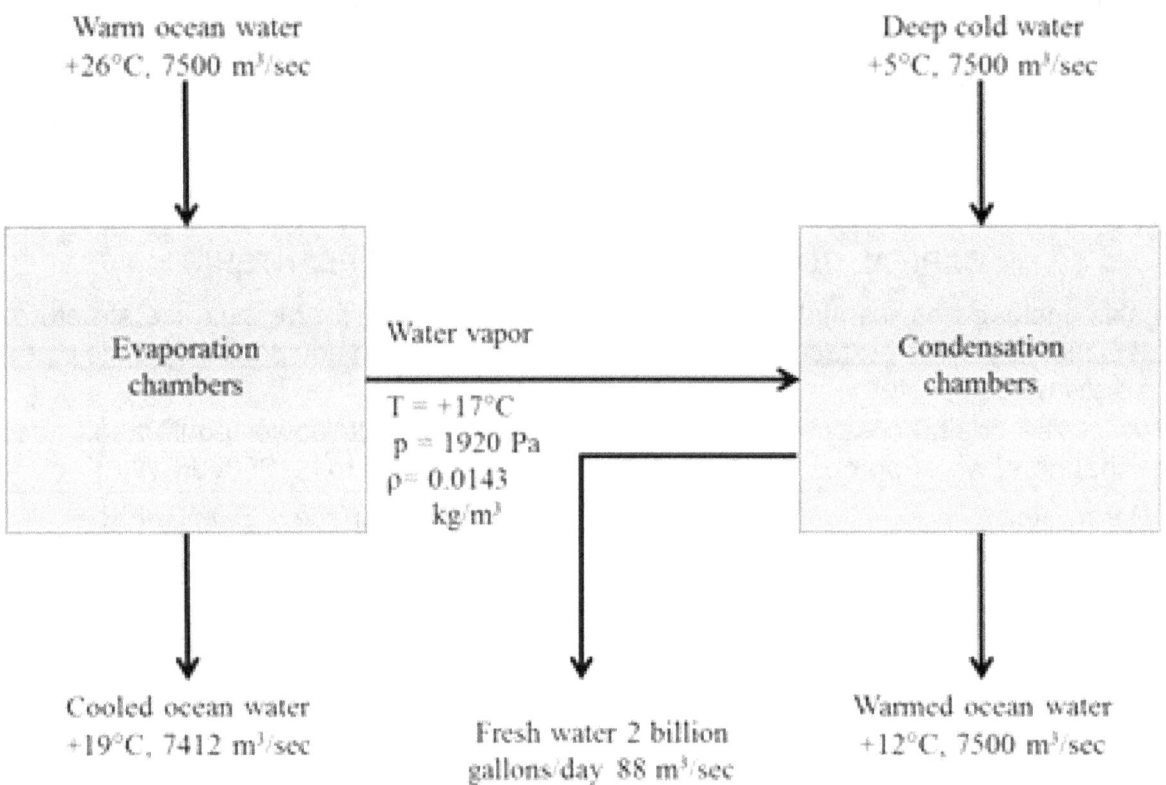

To desalinate ocean water by evaporating warm surface water and condensing the vapor on heat exchangers cooled by deep cost water poses the following major engineering challenges.

- For a fresh water output of 2 Billion gallons per day, 175 Billion gallons per day of warm surface water and 200 Billion gallons per day of deep cold water is required. That's equivalent in volume to flooding Manhattan Island to a depth of 35 feet every 6 hours.
- At 17°C water vapor pressure is only 0.019 atmosphere vapor density is 1/100,000th of liquid water.
- For 2 Billion gallons of fresh water per day, vapor volumetric flow rate is 6.2 million cubic meters per second.
- Flow velocity of low pressure vapor must be low to avoid excessive pressure drop. For 30 meters per second flow velocity and 2 Billion gallons per day of fresh water, vapor flow duct area is 200,000 square meters.

Various approaches have been proposed for evaporating the warm surface area, including flowing warm surface water through heat exchangers that transfer heat to water in a low pressure evaporation chamber and flowing the vapor directly into a water sump inside a low pressure chamber.

OTEC-ICE plantships enables the new, attractive, low cost, high energy efficiency approach illustrated in Figure 6.8. Thin films of warm surface seawater flow down the surfaces of vertical plastic sheets, evaporating on the order of 1% of their volumetric flow into low pressure water vapor.

FIGURE 6.8
EVAPORATION OF WARM OCEAN WATER USING FALLING WATER FILMS ON PLASTIC SHEETS.

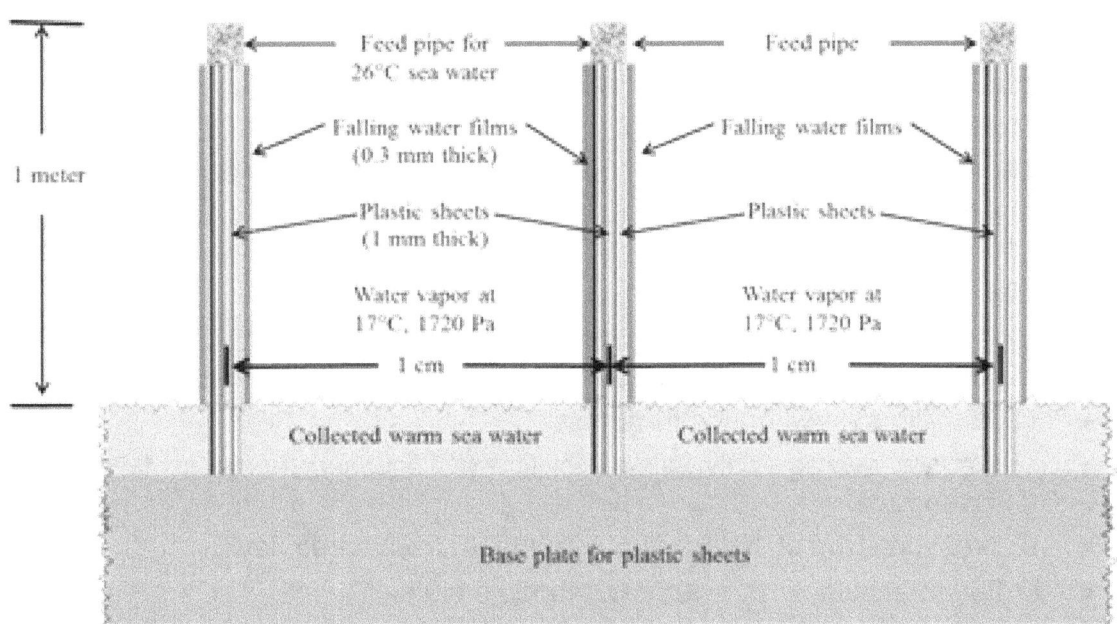

The plastic sheets are located in a large evacuated chamber located inside the OTEC-ICE plantship. The water inlets to the chamber is positioned 10 meters above the ocean, with the one atm pressure difference between the 1 atm air outside of the plantship, and the 0.019 atm pressure in the evacuated chamber providing the force to required to lift the inlet water 10 meters above the ocean surface.

The warm water films flow down the 1 meter virtical height of the plastic sheets and are collected at the bottom of the chamber. The collected water is lower in temperature, due to the evaporation process, with an outlet temperature that is on the order of 17°C, compared to an inlet temperature of about 25°C

The collected water then flows by gravity back to the ocean, with assistance by a pump of approximately 0.1 atm to compensate for the frictional pressure loss when the water flows down the 1 meter vertical plastic sheets.

The water vapor flows out from between the plastic sheets to external floor grids, through which the vapor flows downwards to a second chamber, the condenser chamber, that is located beneath the evaporator chamber as illustrated in Figure 6.9. In the condenser chambers the water vapor is condensed to liquid on heat exchange surfaces cooled by deep cold water, as illustrated in Figure 6.10. Table 6.7 summarizes the evaporator design parameters for 2 Billion gallons of fresh water per day, while Table 6.8 summarizes the condenser design parameters.

TABLE 6.7
SUMMARY OF EVAPORATOR DESIGN PARAMETERS FOR SEA WATER DESALINIZATION

Modular evaporator units
- Module 4 m x 4 m length/width (16 m2 floor area)
- Module height is 1 meter
- 100 vertical plastic sheet/m2 of floor area (1600 sheets per module unit)
- Plastic sheets are 1 millimeter thick, 1 meter high, spaced 1 centimeter apart

Module flow parameters
- 0.2 kg/sec downward water film flow rate on each side of plastic sheet, per meter of sheet length (total of 0.4 kg/sec per meter for both sides)
- 0.383 millimeter water film thickness
- 0.51 m/sec average flow velocity of film (0.75 m/sec maximum flow velocity of film surface)
- ΔP pressure drop of 10^4 Pa (1.5 psi) due to downwards flow
- 12,000 total number of evaporator module units
- 192,000 m^2 total floor area in 12,000 units
- 38 million m^2 total sheet area in 12,000 units (2 sides)
- 7.3 kg/sec fresh water output from each module unit (167,000 gallons per unit per day, 2 billion gallons per day from 12,000 module units)
- 2000 kg weight of 1600 plastic sheets in module unit
- $6000 capital cost of plastic sheets in module unit at 3$/kg
- 0.34 cents per 1000 gallons of fresh water for amortized cost of plastic sheets (30 year amortization period)

TABLE 6.8
SUMMARY OF CONDENSER DESIGN PARAMETERS FOR SEA WATER DESALINIZATION

Modular condenser units
- Module 4 x 4 meters length/width (16 m² floor area)
- Module height is 1.4 meters
- 100 aluminum sheet heat exchangers per m² of floor area, spaced 1 centimeter apart (1600 per modular unit)
- Aluminum sheet heat exchanger has 2 aluminum sheets, each 1 millimeter thick, enclosing a central 2-millimeter-thick water channel 4 meters in length and 1.4 meter high
- 12,000 total number of condenser module units
- 192,000 total floor area of 12,000 units
- 54 million m² total heat exchange area in 12,000 units (2 sides of heat exchangers)

Module flow parameters
- +5°C cold water inlet temperature, +12°C outlet temperature, 7500 m³/sec total flow rate
- 1.56×10^{-3} m³/sec water flow rate per heat exchanger
- 2.8×10^{-3} m² water channel flow area
- 0.56 m/sec water flow velocity
- 10^4 Pa pressure drop for 4 meter long heat exchanger
- 7.8 kg/sec fresh water output from each module unit (167,000 gallons per unit per day, 2 billion gallons per day from 12,000 module units)
- 12.5 tonnes weight of aluminum heat exchanger in module unit
- 450 million dollars capital cost of aluminum sheets in 12,000 module units at 3$/kg
- 1 cent per 1000 gallons of fresh water for amortized cost of aluminum heat exchangers at $3/kg

Figure 6.11 shows a layout of the OTEC-ICE plantship for desalination, while Table 6.9 summarizes its design parameters.

The dimensions of fresh water OTEC-ICE plantship is greater than the power generation OTEC-ICE, 1400 meter length vs 1320 meters and 1080 meters beam vs 870 meters.

Its refrigeration power and cost are also slightly greater, though still very modest, as given in Table 6.10. For the 2000 MW(e) OTEC-ICE plantship, refrigeration power is 5.37 MW(e) at an annual cost of 2.32 million dollars per year. For the 2

Billion gallons of fresh water per day, refrigeration power is 6:7 MW(e) at an annual cost of 2.96 million $ per year.

Table 6.11 summarizes the capital cost components for the 2 Billion gallon per day OTEC-ICE plantship. The capital cost of the plantship itself – insulation, freezing, and cold water pipes – is 364 million dollars. Adding in the 72 million dollars for the evaporation modules and 450 million dollars for the heat exchanger condensate modules, the total capital cost is 886 million dollars. Amortized over a 30 year period that is only 4 cents per 1,000 gallons, much less than 4 dollars per 1,000 gallons, the cost using existing desalination technology.

Transporting the desalinated water from the OTEC-ICE plantships to drought regions, both coastal and inland is another major engineering challenge. Methods for low cost transport of fresh water using ice structures for very low cost ships and ice pipelines are described in the book, The Ice World Cometh.

Finally, it is possible to combine both the electric generation and desalination capabilities into a single OTEC-ICE plantship as illustrated in Figure 6.12. The evaporation/condensation chambers are positioned above the closed ammonia cycle modules.

FIGURE 6.9
LAYOUT OF WATER VAPOR EVAPORATORS AND CONDENSERS.

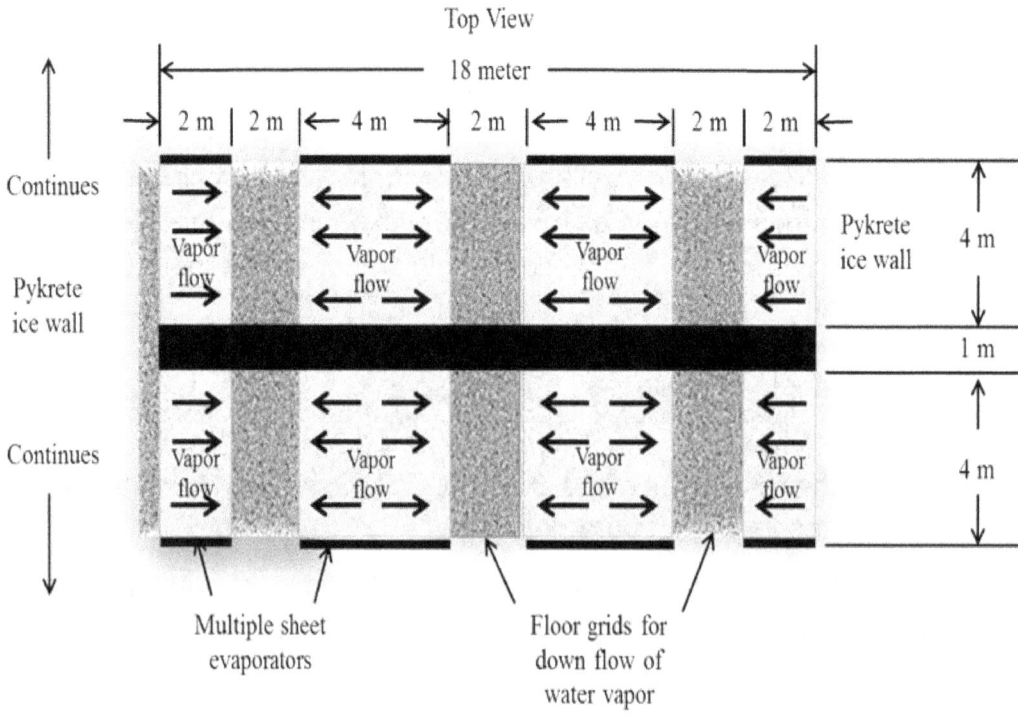

FIGURE 6.10
SIDE VIEW OF HEAT EXCHANGERS FOR CONDENSATION OF WATER VAPOR IN OTEC-ICE LOW PRESSURE CHAMBERS.

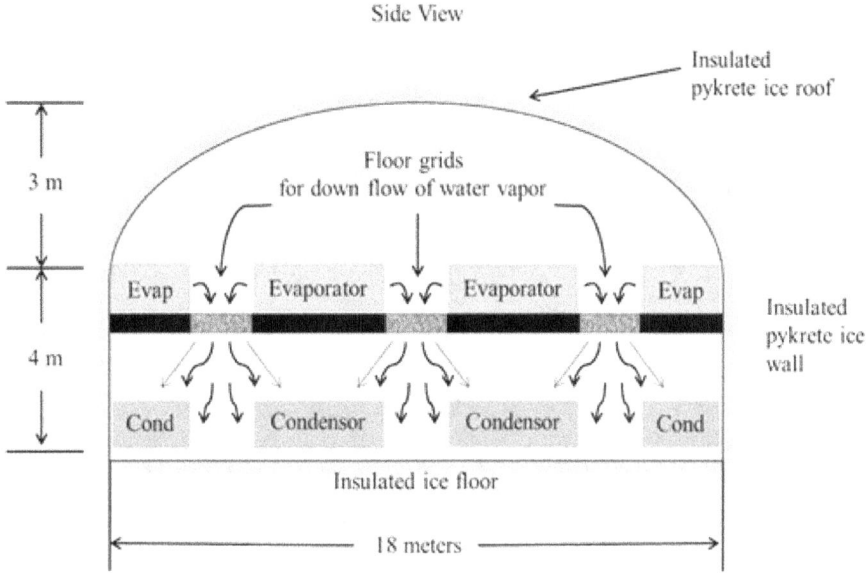

The inlet warm surface water and cold deep water flow first to the closed ammonia cycle module to generate the electric power. Leaving the power generation modules, the warm water now cooled by several degrees and the cold water stream now warmed by several degrees would flow to the evaporator and condenser chambers to evaporate and condense fresh water.

FIGURE 6.11
LAYOUT OF OTEC-ICE PLANTSHIP FOR FRESH WATER PRODUCTION OF 2 BILLION GALLONS PER DAY.

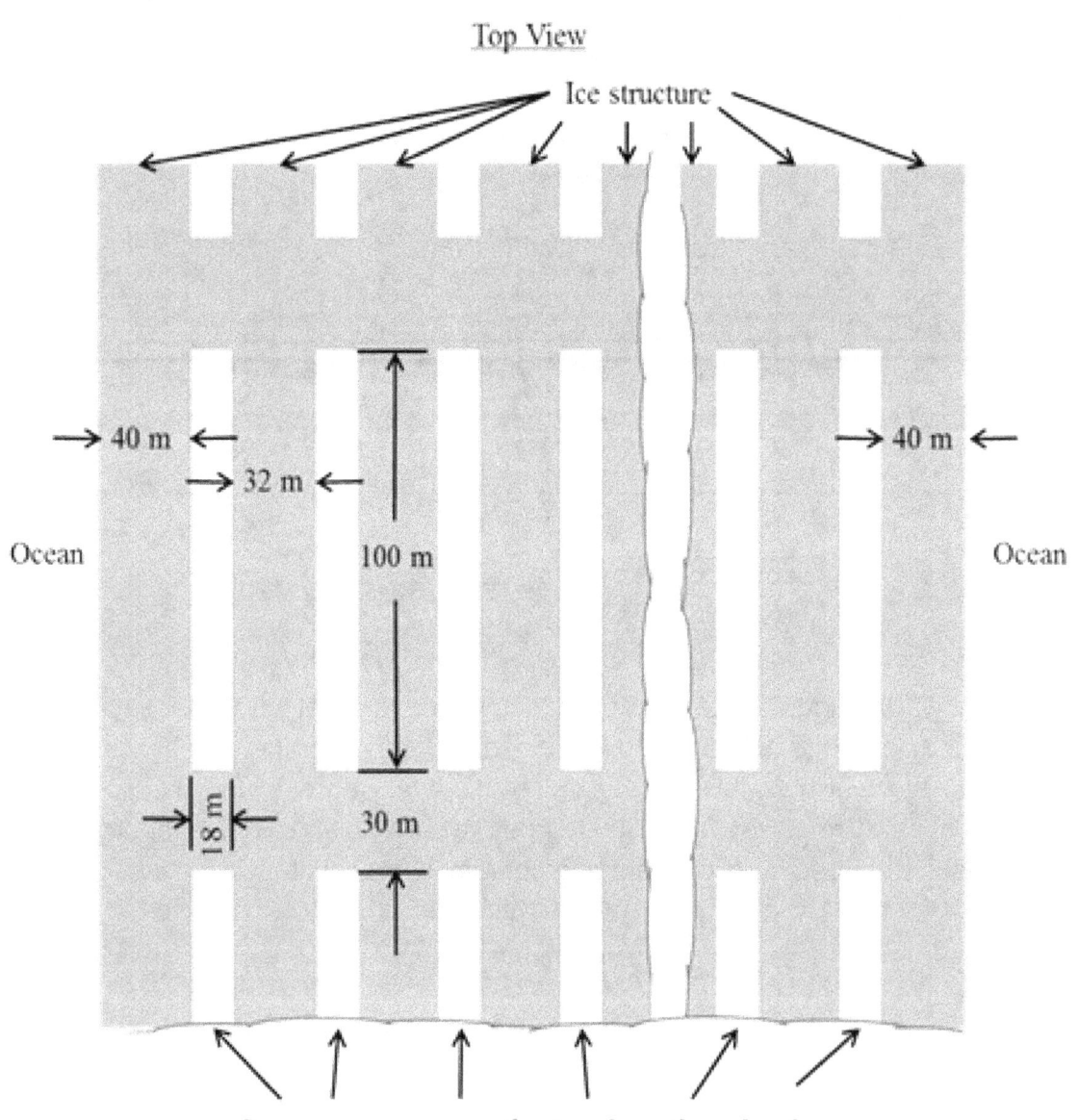

FIGURE 6.11 (CONTINUED)
LAYOUT OF OTEC-ICE PLANTSHIP FOR FRESH WATER PRODUCTION OF 2 BILLION GALLONS PER DAY.

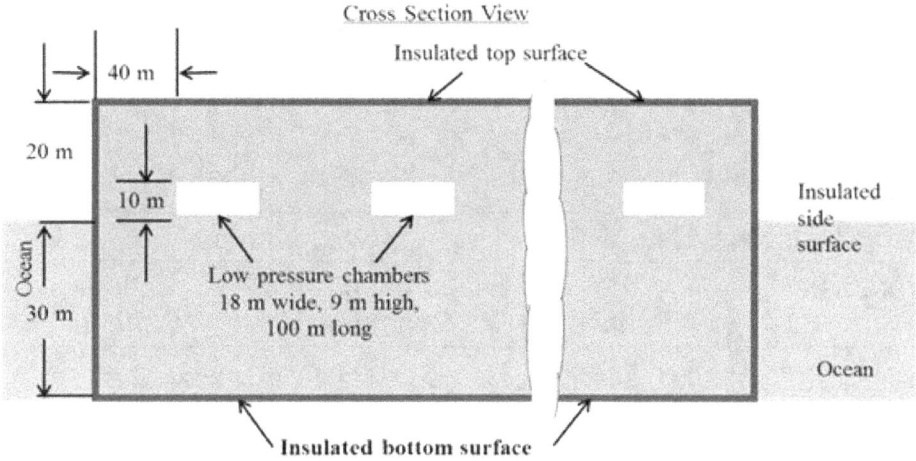

Overall dimensions of OTEC-ICE Fresh Water Plantship
- 1400 meter length
- 1080 meter width
- 20 meter height above ocean surface
- 30 meter depth below ocean surface

TABLE 6.9
SUMMARY OF OTEC-ICE PLANTSHIP DESIGN PARAMETERS FOR FRESH WATER DESALINIZATION

Evaporation/condenser chambers

200 low pressure chambers on OTEC-ICE plantship, 18 meters wide, 100 meters long, 9 meters high

Water vapor in chamber at 17°C, 1720 Pa pressure

60 module evaporator units in each chamber plus 60 module condenser units

3 module units across width of chamber, 20 lines of 3 units along 100 meter length of chamber

32 meter thick ice walls between 18 meter wide adjacent chamber, 30 meter thick ice walls between ends of 100 meter long chambers

Top of evaporator plastic sheet feed points are 10 meters above ocean surface

OTEC-ICE plantships

1400 meter length

1080 meter beam

Top surface of plantship is 20 meters above ocean surface

Bottom surface is 30 meters below ocean surface

TABLE 6.10 REFRIGERATION POWER AND COST FOR -30°C ICE INTERIOR OF 2 BILLION GALLONS OF FRESH WATER PER DAY OTEC-ICE DESALINIZATION PLANTSHIP

Basis: +26°C warm surface water temperature

+5°C deep cold water used as intermediate coolant between +26°C and -30°C

-30°C ice interior insulation has outer temperature of +5°C and inside temperature

of -30°C, except for evaporation chamber insulation, with outer temperature of +17°C

3.49 COP

Plantship Insulation Area				
Parameter	Top Surface	Evaporation/Cond. Chambers	Bottom Surface	Side Surface
Insulation type	Sand/perlite bags	CO_2 gas cell layer	CO_2 gas cell layer	Armored dry sand
Area, m^2	1.51×10^6	1.08×10^6	1.51×10^6	248,000
Thickness, cm	30	30	30	30
k, W/mK	0.08	0.0155	0.0155	0.14
g(th), w/m^2	9.9	1.8	1.8	16.9
P(e), w/m^2	2.84	0.52	0.52	4.67
Total power, MW(e)	4.28	0.56	0.78	1.16
Annual cost, M$/year (5 cents/KWH)	1.87 M$/year	0.24 M$/year	0.34 M$/year	0.51 M$/year
Total power for all surface = 6.78 MW(e)				
Total refrigeration power cost = 2.96 M$/year				

TABLE 6.11 CAPITAL COST OF OTEC-ICE PLANTSHIP PRODUCING TWO BILLION GALLONS PER DAY OF DESALINATED FRESH WATER

Included: Thermal insulation, ice structure freezing, cold water pipes, evaporator and condenser modules

Not included: Water pumps, maintenance, refrigeration and propulsion equipment, and crew quarters

Insulation				
Parameters	Top Surface	Low Pressure Chambers	Bottom Surface	Side Surface
Insulation cost, $/m^2	15.45	15	15	153
Area, m^2	1.51×10^6	1.08×10^6	1.51×10^6	248,000
Capital cost, million $	23.3 M$	16.2 M$	22.6 M$	37.9 M$
Total capital cost for insulation =				100 M$
Total ice volume =				57 million m^3
Capital cost for freezing (@ $2/m^3) =				114 M$
Capital cost of cold water ice pipes (8 pipes each 1.4 km long @ 12.5 M$/km) =				150 M$
12,000 evaporator modules =				72 M$
12,000 condenser modules =				450 M$
Total capital cost =				886 M$
30 year amortized capital cost =				29.5 M$/year
Amortized capital cost/1000 gallons =				4.0 cents/1000 gallons

FIGURE 6.12 ELECTRIC GENERATION AND DESALINATION CAPABILITIES INTO A SINGLE OTEC-ICE PLANTSHIP

Cross Section

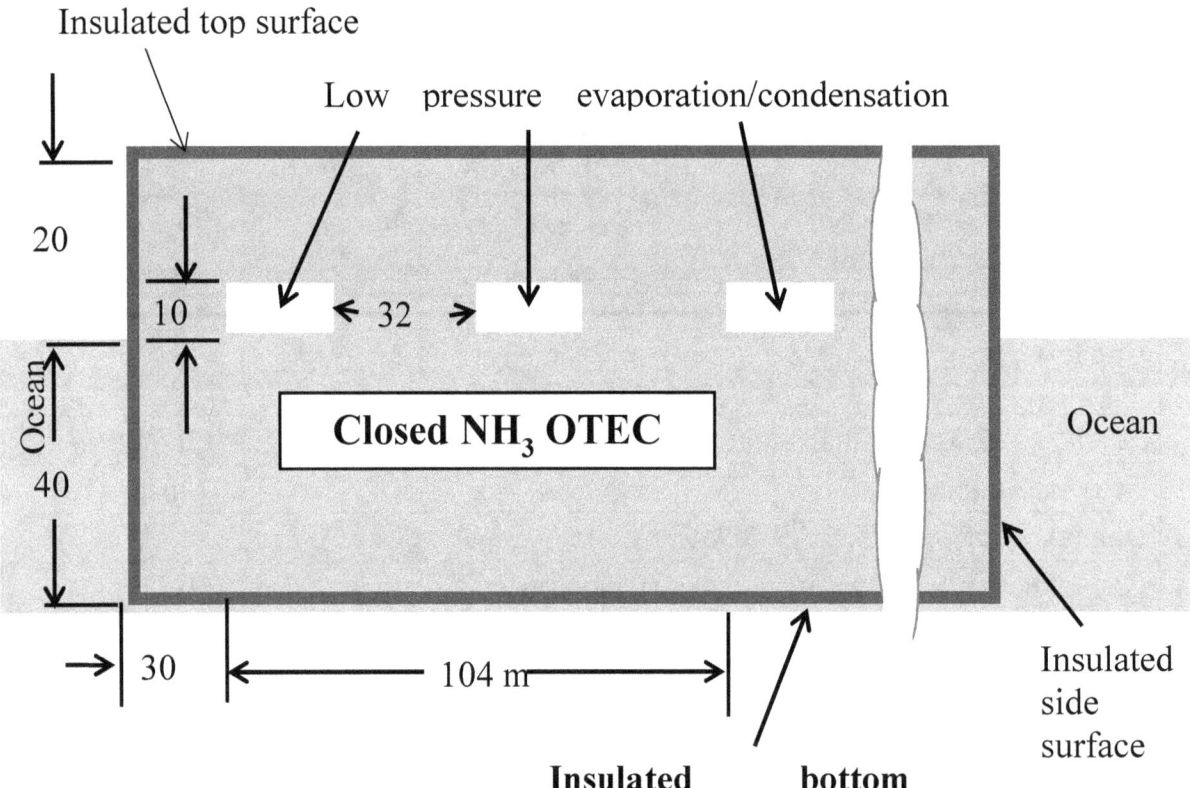

Using a OTEC-ICE plantship, 1400 meters in length and 1080 meters wide, the combined power/fresh water plantship would produce 2000 MW(e) electric power and 1.9 Billion gallons per day of fresh water.

At a revenue price of 2 cents per KWH (e) and 10 cents per 1000 gallons the 30 year revenues would be 12.6 Billion dollars, much greater than the 30 year costs of 1.8 Billion dollars, which include the cost of the plantship, i.e., insulation, freezing, cold water pipe, and refrigeration, evaporator and condenser modules, the closed ammonia cycle power system, propulsion equipment, pumps, crew quarters, personnel, and equipment.

OTEC-ICE technology appears very promising for electric power generation and fresh water production. It uses known materials and processes, requiring only engineering and testing prototype systems to demonstrate its practicality.

Once demonstrated it can be rapidly implemented to help mitigate the serious effects of massive world-wide droughts and also generate massive amounts of low cost electric power.

One OTEC-ICE plantship generating 2000 MW(e) of electric power and 1.9 Billion gallons of fresh water daily can supply 4 million people with the electric power at the global average of 500 watts per person, and 19 million people with fresh water at an average of 100 gallons per day.

100 OTEC-ICE plantships can supply 400 million people, 1/20th of World population, with 500 watts (e) of electric power and 1.9 Billion people, 1/4 of world population, with 100 gallons of fresh water daily.

The total capital cost for 100 OTEC-ICE plantships to serve hundreds of millions of people with electric power and fresh water? 180 Billion dollars amortized over a 30-year period is only 6 Billion dollars per year – much less than power and fresh water from existing sources.

Time to implement 100 OTEC-ICE plantships once the technology is demonstrated? Only a few years. The annual capital investment for implementing 100 plantships in 5 years is only 36 Billion dollars annually – 1/2000th of current annual world GDP.

Clearly, implementation of OTEC-ICE plantships will be extremely rapid, once demonstrated.

List of References:

1. Classic Penod Collapse of the Central Mayan Lowlands; Insight about human-Environment Relationships for Sustainability, B.W. Turner and Jeremy A Sabloff, Proceeding of National Academy of Science 109, 35, 13908 – 13914 (2012)

2. Collapse Mesopotamia; https://www.learner.org/exhibits/collapse

3. Collapse, Jared Diamond

4. Only 60 years of Farming Left If Soil Degradation continues, https://www.scientificamerican.com/articles/only-60-years of-farming-left-if-soil-degradation-continues/

5. https://www.the guardian.com/comment is free/2015/mar/25/treating-soil-like-dirt-fatal-mistake-human-life?CMP=share_btn_link

6. Input constraints to food production: the impact of soil degradation, R.J. Rickson, et al, Food security, 7 Issue 2, Pp.351-364 (April 2015)

7. Projected changes in yields of selected crops with global warming, https://en.wikipedia.org/wiki/file:projected_changes_in_yields_of_selected_crops_with_global_warming.jpg Author, Enescot

8. http://weather.com/science/environment/news/california-historic-drought-world-brazil-africa-korea

9. news.nationalpost.com/news/world/the-water-table-table-is-dropping-all-over-the-world-new-nasa-study-reveals-global-drought

10. https://en.wikipedia.org/wiki/desalination

11. http://en.wikipedia.org/wiki/water_in_california

12. The Worlds Water, Volume 7, Peter Gleick Pacific Institute (2009)

13. Ocean Thermal Energy Conversion, Lu is A Vega, Encyclopedia of Sustainability Science and Technology, Springer, August 2012, Pp.7296-7328, hinmrec.hnei.hawaii.edu/wp,,,Article-OTEC-by-Vega-Aig-2012.pdf

14. First Generation 50 MW OTEC Plantship for the Production of Electricity and Desalinated Water, Luis A Vega and Dominic Michaelis, OTC 20957, 2010 Off-Shore Technology Conference, Houston, Texas, USA, 3-6, May 2010

List of Figure Credits

Figure 6.1 Projected changes in yields of selected crops with global warming, https://wikimedia.org/wiki/file

Figure 6.2 Map of California Drought Status, https://commons. Wikimedia.org wiki/file: California_Drought_status_October 21_2014.png. Author, Michael Brewer, NCDC/NOAA

Figure 6.3 Map of West U.S. Drought, https://drought monitor.Unl.edu/home/regional drought monitor.as ...Author, Anthony Artuso, NOAA/NWS/NCEP/CPE

Figure 6.4 Desalination of sea water using ocean thermal Energy. Author, Powell

Figure 6.5 Typical Large Ice Island Structure, Author, Powell

Figure 6.6 Arrangement of 50 MW(e) units on 2000 MW(e) OTEC-ICE Plantship, Author, Powell

Figure 6.7 Flow Sheet for OTEC-ICE Plantship to Produce 2 Billion Gallons of Fresh Water Daily, Author, Powell

Figure 6.8 Evaporation of warm ocean water using falling water films as plastic sheets, Author, Powell

Figure 6.9 Layout of water vapor evaporators and condensers, Author Powell

Figure 6.10 Side view of Heat Exchangers for Condensation of water vapor in OTEC-ICE Low Pressure Chambers, Author, Powell

Figure 6.11 Layout of OTEC-ICE Plantship for fresh water production of 2 Billion Gallons per Day, Author, Powell

Figure 6.12 Cross Section View, Author, Powell

Chapter 7

Political and Economic Barriers Against Saving Spaceship Earth

There are great technology and economic challenges if we are to save Spaceship Earth. We must transition from fossil fuels, develop practical, clean new sustainable energy sources, find ways to keep atmospheric carbon dioxide concentrations at safe levels and insure sufficient agricultural capability, soil and water to continue to feed Billions of humans.

Great as the above challenges are, they are dwarfed by the challenge posed by our inherent human nature. Humans are inherently tribal, ready and willing to exploit others, pollute and destroy the environment for personal gain, and don't worry about the Deluge that comes after them.

Judging from a human perspective these features have not been detrimental to human progress, in fact, in the minds of many, they have been advantageous. As described in Chapter 1, in 70,000 years, human population has grown from 7,000 people to 7 Billion individuals. We are no longer lope through forests and savannahs, hoping to find some food – a few nuts or fruits, a squirrel or rabbit. Instead, we grow amazing amounts of reliable, tasty, and healthy food, travel thousands of miles when and where we want, have large comfortable homes, communicate worldwide using radio, TV, computers, cellphones, magazines, newspapers and books, live much longer lives, conquer diseases, etc., etc.

Judged from the perspective of the rest of the Earth, however, humanity's success story has come at the expense of much of the other life forms on Earth. Humans have already wiped out many species. By 2100, scientists project that humans will have wiped out 1/2 of the species on Earth, the 6th mass extinction, joining the 5 previous caused by asteroids and volcanic eruptions.

In addition, by 2100, the oceans will have acidified to the point where the world's coral reefs will have died, along with all the ocean's shelled creatures. Much of the World's forested areas will be gone, turned into deserts.

All of the above wil happen, if humanity continues on its present path, leading to the disastrous crash of Spaceship Earth and the possible extinction of humanity itself.

So, why isn't humanity doing everything in its power to prevent the crash of Spaceship Earth? The reason? Humanity's inherent nature to form tribes, fight the other tribes, exploit each other in a given tribe for personal gain, pollute the environment and think only about the present and themselves, and not about future generations.

Examples? Corporations, investors, workers involved in the production, sale, and profits of fossil fuels – oil, coal, and gas – lobby against the development and implementation of renewable energy sources, fund politicians and climate deniers and fight regulations on greenhouse gas emissions.

Concern about the World's environment 50 years from now? Who really cares when most people live with a host of today's problems? Is my job at risk? I can't afford health insurance? My car just broke down – I don't have money to fix it? My savings are not enough to retire? I can't affort to send my children to college?

Spending money today to ensure a sustainable world 50 year from now is a hard sell for most people. Fortunately, a significant fraction of humanity does want to achieve a sustainable Earth. New renewable energy technologies, e.g. wind and solar power are being implemented. Technologies to capture greenhouse gases from power plants and the atmosphere are being developed. Carbon taxes and credits are being formulated.

In developing new technologies for Spaceship Earth, it is very important that they offer economic and social benefits relatively quickly, and minimize social and economic disruptions. For example, beamed power from space solar power satellites, described in Chapter 4, offers limitless worldwide electric power at much lower cost than from existing power plants, and could begin operation in only a few years.

Implementing beamed space solar power will cause some economic and social disruptions. Existing power plants will shut down. Fossil fuel production will drop, affecting investors and workers. However, the full implementation of beamed space solar power will occur over several decades, with any disruptions more than compensated for by the greater world GDP and the millions of new jobs that will be created by thousands of new industries that lower cost electric power will provide.

A similar situation holds for the other 2 new technologies described in Chapter 5, extracting CO_2, and Chapter 6, generating massive amounts of desalinated low-cost fresh water.

Extracting carbon dioxide from the atmosphere will enable major world-wide economic and social benefits, smaller drought areas, less severe storms, surges, lower sea levels, less coastal flooding, fewer crop failures, less heat wave deaths, etc.

The cost of CO_2 extraction appears affordable in terms of World GDP, and will provide millions of new construction and operating jobs.

Generating massive amounts of low-cost desalinated fresh water using the new technology described in Chapter 6 will also have major World-wide economic and social benefits, enabling sustainable, lower cost food supplies and large numbers of new jobs for farmers, processors, transport workers and retailers.

To be effective and avoid the crash of Spaceship Earth, the whole World must united and cooperate in developing and implementing these 3 new technologies.

Action by only one of the World's tribes, read "Nation", to carry out one or more of these new technologies will not avoid a crash. All the world's tribes, must cooperate in good faith and not try to take advantage of each other.

Can they unite and really cooperate? That is the hardest and ultimate challenge. Hopefully they will.

*"**The Raft of the Medusa** (<u>French</u>: *Le Radeau de la Méduse* is an oil painting of 1818–1819 by the French Romantic painter and lithographer Théodore Géricault (1791–1824). Completed when the artist was 27, the work has become an icon of French Romanticism. At 491 cm × 716 cm (16' 1" × 23' 6"), it is an over-life-size painting that depicts a moment from the aftermath of the wreck of the French naval frigate *Méduse*, which ran aground off the coast of today's Mauritania on 2 July 1816. On 5 July 1816, at least 147 people were set adrift on a hurriedly constructed raft; all but 15 died in the 13 days before their rescue, and those who survived endured starvation and dehydration and practiced cannibalism. The event became an international scandal, in part because its cause was widely attributed to the incompetence of the French captain.

Géricault chose to depict this event in order to launch his career with a large-scale uncommissioned work on a subject that had already generated great public interest. The event fascinated him, and before he began work on the final painting, he undertook extensive research and produced many preparatory sketches. He interviewed two of the survivors and constructed a detailed scale model of the raft. He visited hospitals and morgues where he could view, first-hand, the colour and texture of the flesh of the dying and dead. As he had anticipated, the painting proved highly controversial at its first appearance in the 1819 Paris Salon, attracting passionate praise and condemnation in equal measure." From https://en.wikipedia.org/wiki/The_Raft_of_the_Medusa

Chapter 8

How Much Time Do We Have to Prevent Spaceship Earth from Crashing?

Time is very short. If humanity continues on its "business as usual" emissions path, relying on fossil fuels to support a growing world economy, we will reach an atmospheric carbon dioxide concentration of 800 ppm, from today's 400 ppm, by 2060.(1)

The resultant global temperature increase since 2000 AD will be 2 degrees Centigrade (3.6 degrees Fahrenheit) by 2060.(1) If business as usual emissions were to continue, by 2100 AD, global temperature would increase by 4 degrees centigrade (7.2 degrees Fahrenheit).

The effects of temperature rise increase rapidly with the magnitude of the temperature rise. For example, Figure 8.1 shows the projected changes in yields of 5 selected crops – Africa maize, Asia rice, India wheat, US Maize, and US soybean – as a function of temperature increase.(2)

With a 2 degree centigrade global temperature rise, the average yield for the 5 crops would decrease on average about 20%; with a 4 degree centigrade rise, the average crop yield would drop by 40%. If this occurs, there will not be enough food to feed the projected population of 9 Billion humans. We must transition from fossil fuels to clean, renewable energy sources to stop global warming and do it quickly.

How long will it take to develop, test, and begin the implementation of the 3 new global technologies described in this book?

Beamed power from space solar power satellites

Extraction of carbon dioxide from the atmosphere

Massive, low cost desalinization of ocean water

To be conservative, we project that each of the 3 new technologies will take on the order of 10 years to develop and test before full-scale operating systems can begin to be implemented.

Faster implementation times are possible, as demonstrated historically with the Manhattan and Apollo projects. Starting from virtually no nuclear technology, the Manhattan project achieved nuclear weapons in only 4 years. The Apollo project landed astronauts on the Moon in 7 years after President Kennedy announced the program.

Major world-wide implementation of the 3 new technologies over the next 20 years following construction of the initial systems. Including 10 years for development by 2047, 30 years from now, a major portion of the World's electric power will be beamed from space power satellites, CO_2 extraction facilities will be extracting Billions of tons of carbon dioxide from the atmosphere, and hundreds of Billions of desalinated ocean water will be produced daily.

Clearly, development of these projects will have to be initiated by major industrial nations, the US, China, UK, European Union, and Russia – but they should be carried out as international projects with access to and involvement by other countries. The economic and social benefits to all the World's countries and populations should be strongly stressed.

FIGURE 8.1
PROJECTED CHANGES IN YIELDS OF SELECTED CROPS WITH GLOBAL WARMING

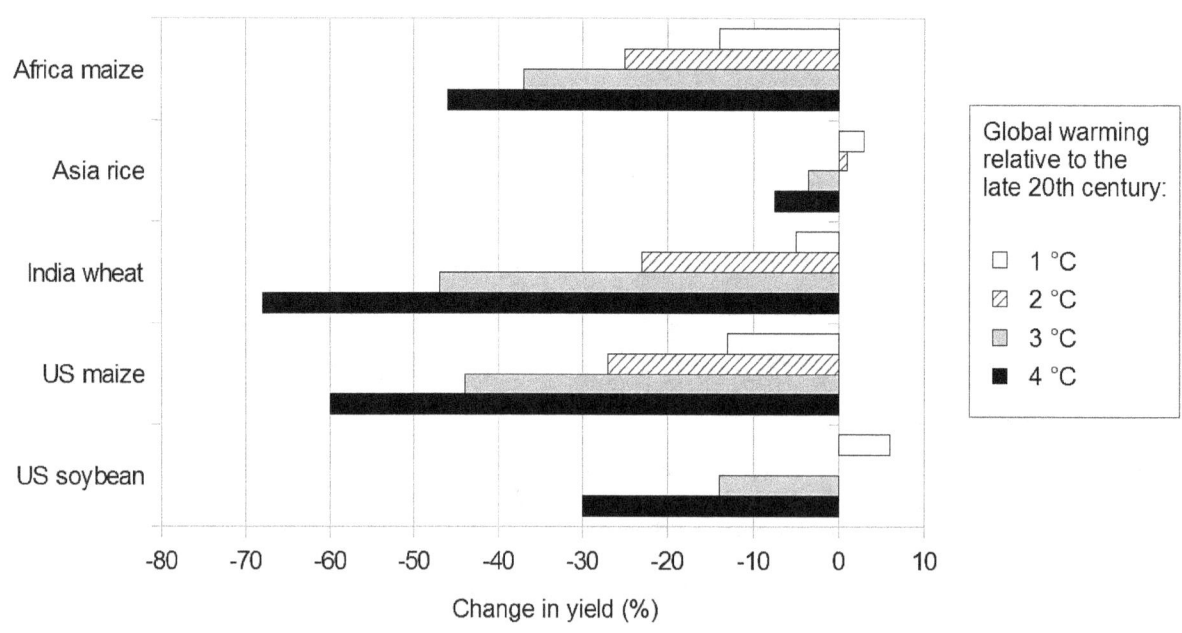

List of References

1. Projected Atmospheric Greenhouse Gas Emissions https://www.epa.gov/climate-change-science/

2. https://en.wikipedia.org/wiki/climate_change_and_agriculture

List of Figure Credits

Figure 8.1. Projected changes in yields of selected crops with global warming, https://en.wikipedia.org/wiki/file:projected_changes_in_yields_of_selected_crops_with_global_warming.jpg Author, Enescot

Appendix A

Removal of Carbon Dioxide from Earth's Atmosphere

The basic process for extraction of carbon dioxide (CO_2) from the atmosphere is illustrated in the basic flowsheet shown in Figure A-1. Atmospheric air flows through packed beds of small particles that absorb most of the CO_2 present in the atmosphere. Flow through the particle beds continues until the particles become saturated with CO_2.

The particle beds are then closed off from contact with the atmosphere and heated to an elevated temperature to desorb the CO_2 from the particles, forming a stream of pure, pressurized CO_2.

The extracted pressurized CO_2 is then pumped into an underground rock formation, when it reacts with minerals in the rock to form geologically, solid compounds that isolate CO_2 from Earth's atmosphere for millions of years.

After desorption of the CO_2 absorbed in the particle beds, they cool down and opened for flowing atmospheric air, to absorb a new load of CO_2 from the atmosphere.

Typical cycle times for the process – absorption followed by more absorption – are on the order of several days. The process would be carried out by hundreds of independently operating sites around the world, with a total CO_2 extraction rate on the order of 30 Billion tonnes per year, comparable to the amount presently being released to the atmosphere by the combustion of fossil fuels.

FIGURE A-1
PROCESS FOR ABSORBING CARBON DIOXIDE FROM THE ATMOSPHERE USING PACKED BEDS OF ABSORBER PARTICLES.

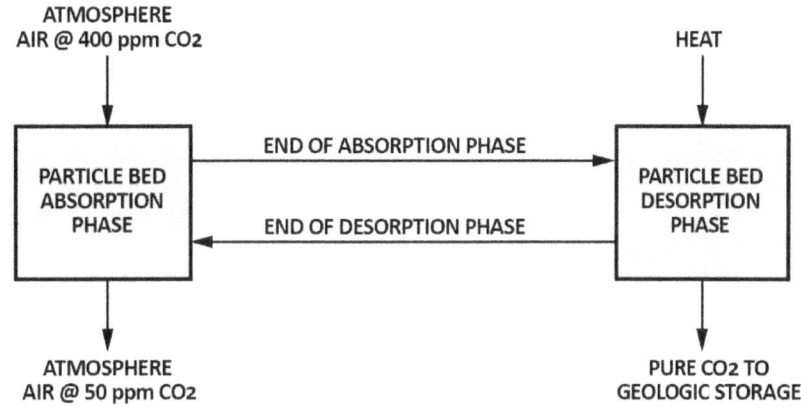

Figure A-2 illustrates the geometry of the atmospheric air flow through a section of particle bed. The particles are packed into a rectangular slab between 2 porous screens that hold the particles, but let air flow into the packed bed on one side and out of the other side. The slab area is typically large, on the order of tens of square meters, with multiple slabs arranged in an array at the extraction site.

After the absorber particles in an assembly of multiple slabs become saturated with CO_2, the assembly is closed off from atmospheric airflow then the particle beds are heated to desorb their CO_2 content. The collected CO_2 is then pumped into underground rock formations where it reacts to form geologically stable materials.

After desorption of the CO_2 from the particle bed, it is again opened to the atmosphere for the next absorption cycle.

There is a small pressure drop for the air flowing through the bed. To compensate for this pressure drop, the incoming airflow is slightly pressurized using conventional industrial fans similar to those used in existing power plants.

FIGURE A-2
FLOW PATH OF ATMOSPHERIC AIR THROUGH PACKED ABSORBER BED

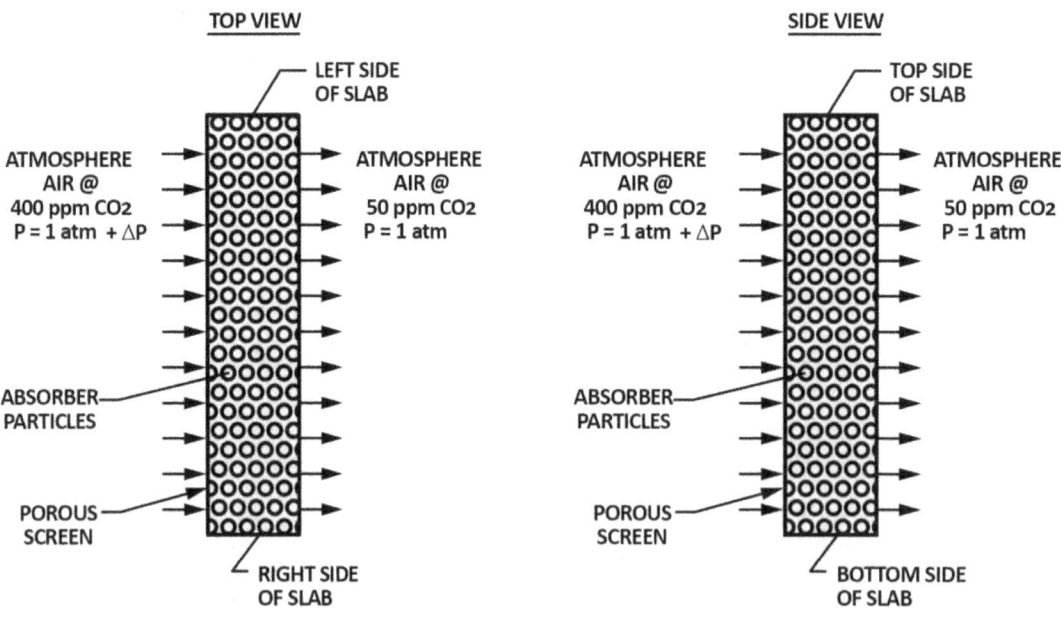

The multiple slabs of absorber particles are arranged in an array as shown in Figure A-3. The slabs are positioned in parallel, with inlet and outlet air flow channels between the slabs. Atmospheric air at 400 ppm CO_2 flows inwards through 1/2 of the flow channels, and outwards through the other 1/2 of the flow channels.

As the atmospheric air flows through an inlet channel, its velocity is highest at the inlet point, and decreases as it moves along the channel, with airflow being diverted to flow through the absorber slabs on each side of the inlet channel. The end of the inlet flow channel is closed, with zero airflow.

Similarly the air outflow channel receives the air flowing out of the 2 absorber slabs on its left and right side, and discharges it back into the atmosphere. The top end of the outflow channel is closed, so that no air from the atmosphere flows into it.

Before describing the design, construction and projected cost of a full-scale CO_2 extraction facility, we first look at the scale of required airflow and the optimum parameters for the absorber slabs.

For an extraction rate of 30 Billion tonnes (1 tonne = 1,000 kilograms) per year from the atmosphere with an inlet concentration of CO_2 of 400 ppm and an outlet concentration of 50 ppm an immense air flow rate will be required. At standard temperature and pressure (sea level and 15 degrees centigrade) the density of atmospheric air is 1.23 kilograms per cubic meter.

FIGURE A-3
ARRANGEMENT OF MULTIPLE ABSORBER SLABS WITH INFLOW AND OUTFLOW AIR CHANNELS.

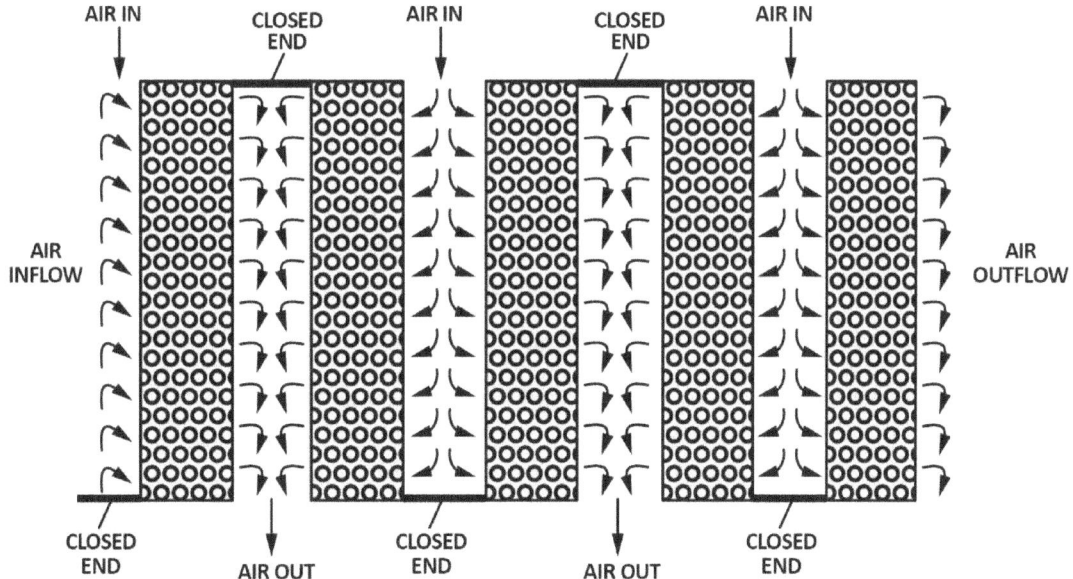

At 400 ppm, a cubic meter of STP air contains:

$m_{(CO2)} = 400 \times 10^{-6} \times 1.23 \times 10^3 \times \left(\frac{CO2\ molecular\ weight}{air\ molecular\ weight}\right)$

$m_{(CO2)} = 400 \times 10^{-6} \times 1.23 \times 10^3 \times \left(\frac{44}{29}\right)$

$m_{(CO2)} = 0.75$ grams CO_2 per cubic meter of air

Reducing the CO_2 concentration to 50 ppm by absorption in particle beds results in an extraction capability of:

$m_{(CO2)} = 0.75 \times 350/400 = 0.66$ grams CO_2 extracted per cubic meter of processed air

To extract 30 Billion tonnes per year of CO_2 from the atmosphere will require an air flow rate of

V_{air}, cubic meters per second.

V_{air} = (CO2 extraction rate in grams/sec)/(grams CO_2 extracted per m³)m³/sec

30 Billion tonnes of CO_2 equals 3×10^{16} grams per year 1 year equals 8760 hours x 3600 seconds/hour extraction equals $3.15 \times 10^7 = 0.95 \times 10^9$ grams/sec and the corresponding air flow rate is then

$m_{(CO2)} = 3 \times 10^{16} / 3.15 \times 10^7 = 0.95 \times 10^9$ grams/sec

$V_{air} = 0.95 \times 10^9$ gms/sec/0.66 grams/m³ – 1.4×10^9 cubic meters per second

1.4 Billion cubic meters of air per second through CO_2 extraction facilities. That's a tremendous air flow. At 10 meters per second air velocity, that corresponds to 140 square kilometers of air flow area for CO_2 extraction.

But the CO_2 extraction air flow area is dwarfed by the air flow area through the wind turbines already operating in the world. There are currently 314,000 wind turbines in the world.[1] With an average rotor diameter of 120 meters, the total air flow area through the 314,000 wind turbines is 3.5 billion square meters, 3,500 square kilometers.

The wind turbine air flow area is already 20 times greater than the CO_2 extraction area. Moreover, it will greatly increase in the years ahead. Today, wind turbines generate 3.7% of the world's total electric power of 21,000 Terawatt Hours.[1] To generate today's world electric power, there would have to be 8 million wind turbines, more than 25 times the present number. The wind turbine air flow area would be more than 500 times greater than the CO_2 extraction flow area. By 2040, world electric power will more than double, so that wind turbine flow area would be more than 1000 times the CO_2 extraction flow area.

Accordingly, while the CO_2 extraction flow area is very large, the flow area for wind turbines, even today, is much greater, and will be much greater still in the decades ahead.

The next issue to address is the design of the absorber slabs – their optimum thickness, what kind of absorber particles would be used, the optimum air flow velocity in through the slab, the airflow pressure drop, the dimensions of the absorber slabs, the spacing between them, etc.

Various kinds of absorber particles have been experimented on, with varying results. For this study, we use the PEI particles tested by Zhang, et al, in their paper, *Capturing CO_2 From Ambient Air Using – Polyethyleneimine – Silica Absorbent In Fluidized Beds*.(2)

The PEI particles are small, several hundred microns in diameter, consisting of inorganic mesoporous silica particles coated with PEI phenethylene that absorbs CO_2 from the atmospheric air. The density of a settled bed of PEI particles is 700 kg/m³, 40% of which is PEI and 60% mesoporous silica. For 250 micron diameter particles, the BET area of the mesoporous silica particle is 250 m²/g, with a pore volume of 1.7 cm³/g and a mean pore diameter of 20 nanometers.

In tests with atmospheric air flowing through a settled particle bed, Zhang and colleagues found that an air/particle bed contact time of 7.5 seconds would remove almost all of the 400 ppm of CO_2 from the in-flowing air. The PEI particles saturated at 5% CO_2 content by mass: i.e., 1 kilogram of PEI particles would hold 50 grams of CO_2 when saturated.

Figure A-4 shows a section of absorber slab with the following variables:

V_A = Superficial flow velocity of air through bed, cm/sec

X_0 = Thickness of bed, cm

ΔP = Pressure drop of airflow through bed, Pascals.

X_0 and V_A are related by the conditions that the air/particles contact time in 7.5 seconds. That is,

Δt = 7.5 seconds = X_0/V_A

For example, for a bed thickness of 30 centimeters (0.3 meters), and Δt = 7.5 seconds, the superficial flow velocity would be 4 centimeters per second.

FIGURE A-4
ABSORBER BED GEOMETRY FOR OPTIMIZATION OF OPERATING PARAMETERS

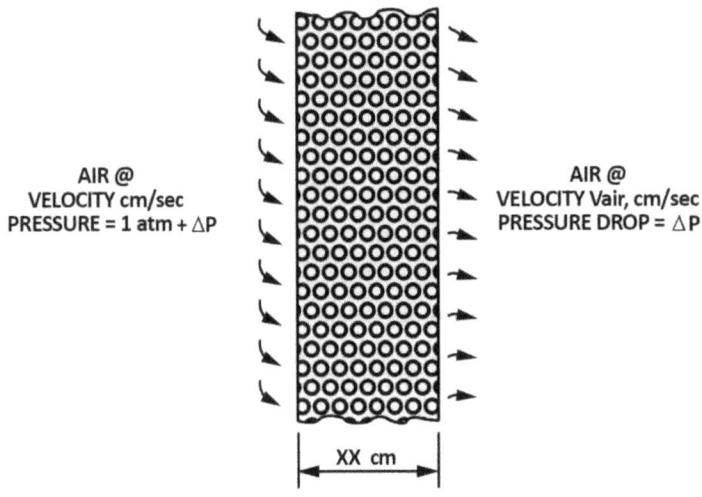

The air pressure drop through the absorber particle bed can be calculated from the Ergun equation, the results of which agree closely with experiments on flow through particle beds.

For a settled particle bed of 500 micron particles at 50 percent of full particle density (50 percent voidage) Table A-1 gives the ΔP per centimeter of bed thickness as a function of superficial velocity V_A, the corresponding thickness of the bed.

TABLE A-1

V_A, cm/sec	ΔP, Pascals/cm of thickness	Bed Thickness, X_o, cm	ΔP Across Bed, Pascals
2	12	15	180
5	31	37.5	1160
10	60	75	4500
20	134	150	21,000

Figure A-5 shows a graph of bed thickness and ΔP across the bed as a function of the superficial velocity V_A. Bed thickness is linear with superficial velocity particle pressure drop across the be roughly scales with $(V_A)^2$.

FIGURE A-5

ABSORBER THICKNESS AND PRESSURE DROP ACROSS IT AS A FUNCTION OF SUPERFICIAL FLOW VELOCITY

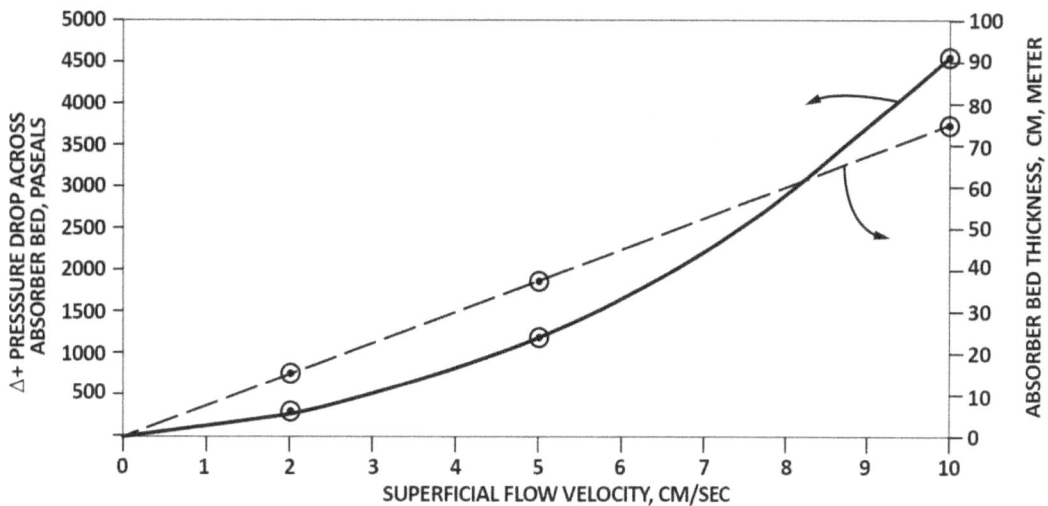

As discussed previously, to remove 30 Billion tonnes of CO_2 from the atmosphere per year, a flow rate of 1.4 Billion cubic meters per second is necessary. The required absorber slab area, particle bed volume, and electric power to overcome the ΔP through the absorber slabs as a function of bed thickness is shown in Table A-2.

TABLE A-2

ABSORBER SLAB AREA, VOLUME, AND AIR FLOW PUMP POWER FOR 1.4 BILLION CUBIC METERS/SEC AIR FLOW RATE

Bed Thickness (cm)	Superficial Velocity, VA cm/sec	Absorber Slab Area (M2)	Particle Bed Volume (M3)	ΔP (Pascals)	Power Gigawatts (e)
10	1.3	1.05×10^{11}	1.05×10^{10}	200	280
20	2.7	5.2×10^{10}	1.05×10^{10}	450	630
30	4	3.5×10^{10}	1.05×10^{10}	720	1000
40	5.3	2.6×10^{10}	1.05×10^{10}	1300	1820
50	6.7	2.1×10^{10}	1.05×10^{10}	2000	2800

Based on the above calculations, the optimum absorber slab thickness appears to be on the order of 30 centimeters. The pumping power for thicker slabs is excessive, while the slab area for thinner slabs is too great. The volume of absorber particles does not depend on slab thickness.

However, the required area, volume, and power are extremely large, even for the optimized slab thickness of 30 centimeters.

Total World electric power generation today is 2,500 Gigawatts, with a projected increase to 4,200 gigawatts by 2040. Assuming that the CO_2 removal from the atmosphere was in operation in 2040, its electric power would increase World electric power generation from 4,200 Gigawatts by 1,000 GW(e) to a total of 5,200 GW(e).

This 25% increase in power generation appears practical, with beamed power from orbiting solar satellites at a cost of 2 cents per KWH(e), as described in Chapter 4, the cost of the 1000 GW(e) power used for the CO_2 atmospheric removal program, would be 175 Billion dollars per year.

Compared to the Trillions of dollars spent annually around the world on infrastructure projects – highways, airports, railroads, seaports, etc. the power cost for CO_2 removal from the atmosphere will be small.

Table A-2 shows the total absorber slab area and volume for a slab thickness of 30 centimeters. The slabs will be arranged inside a large containment structure, into which atmospheric air enters at a CO_2 concentration of 400 ppm, and leaves at a CO_2 concentration of 50 ppm.

As illustrated in Figure A-3 the absorber slabs are positioned parallel to each other, with an inlet airflow channel on one side of the slab and an air outflow channel on the other side. With an absorber slab thickness of 30 centimeters, and 20 centimeters wide inflow and outflow channels, 2 absorber slabs with their inflow and outflow channels, 2 absorber slabs with their inflow and outflow airflow channels have a unit width of 1 meter.

Total slab surface area for 30 centimeter slabs is 3.5×10^{10} square meters. With 2 slabs per meter of width, the assembly volume to contain the 3.5×10^{10} square meters is:

$V_{assembly} = 3.5 \times 10^{10}$ m^2/2 $= 1.75 \times 10^{10}$ cubic meters

A very large total assembly volume, for the CO_2 extraction project.

However, it is small compared to the total construction volume in the world. For example, the Boeing factory in Everett, Washington has a volume of 13.3 million cubic meters, with a footprint of 398,000 square meters. 1,000 Boeing size CO_2 extraction facilities would roughly equal the total volume required for CO_2 extraction from the atmosphere.

Worldwide, there are 1.5 Billon households(4) not including hotels, rooming houses, offices, lodging houses, institutions and camps. At an average of 400 cubic meters per household, 1.5 Billion households occupy 600 Billion cubic meters. Adding in the other construction, total present world construction volume is considerably more than 1000 billion cubic meters, and more than 100 times the assembly volume for the CO_2 project.

Moreover, construction of the CO_2 extraction facilities will be much simpler and less complex than the construction of houses, hotels, large office buildings, etc. Basically, aircraft hanger type buildings with an assembly of absorber slabs inside.

The cycle time for the absorber slabs, CO_2 absorption time plus desorption is determined by the superficial air velocity through the slab, slab thickness, and CO_2 saturation content.

For PEI absorber particles in the 5% by weight CO_2 saturation content, and a settled particle bed density, of 700 kg per m^3, the CO_2 content per square meter of slab area for 30 centimeter thick slabs is:

$(M_{CO2})_{SAT}$ = (thickness, m)(bed density, kg/m^3)(0.05)

$= 0.3 \times 700 \times 0.05 = 11.6$ kilograms/m^2

At the superficial velocity of 4 centimeters per second through the absorber slab, and 0.66 grams (6.6 x 10^{-4} kg) of CO_2 is extracted from each cubic meter of air, the extraction time to reach the CO_2 extraction point of 5% by weight in the absorber particles with a flow rate of 0.04m^3/sec through the slab is

Δt_{SAT} = 11.6/0.04 x 6.6 x 10^{-4}

Δt_{SAT} = 4.4 x 10^5 seconds = 5.1 Days

The CO_2 desorption time is much shorter, on the order of a day. Counting setup time, cycle time, absorption plus desorption, will be on the order of 7 days, 1 week.

Over 1 year of operation, a 1 square meter, 30 cm thick absorber bed would undergo

cycles/year = 365/7 = 52 cycles

With 11.6 kilograms of CO_2 extracted per cycle

M_{CO2} per year+52 x 11.6 = 600 kg per year

Over a 30-year period, total CO_2 extraction

M_{CO2} for 30 years = 18 tonnes CO_2.

The 30-cm thick, 1 square meter absorber slab contains

$M_{absorber}$ = 700 kg/m^3 x 0.3m = 0.21 tonne absorber

The ratio of CO_2 mass absorbed to absorber mass is

$M_{CO2}/M_{absorber}$ = 18/0.21 = 86

The cost of the absorber particles will probably be in the range of $1000 to $2000 per tonne. At $1000 per tonne, the corresponding cost per tonne of CO_2 extracted would be $1000/86 = $11.6 per tonne of CO_2 extracted. At $2000 per tonne of absorber, the corresponding CO_2 cost would be $23 per tonne.

Achieving a goal of $100 per tonne of CO_2 extracted corresponds to a total of 3 Trillion dollars per year for extracting 30 Billion tonnes of CO_2 annually extracted from the atmosphere. For a projected World GDP of more than 200 Trillion dollars per year in 2050. This appears practical if it can avert global environmental catastrophe.

The various CO_2 extraction facilities will have an assembly of absorber slabs inside a hangar type structure, with fresh atmospheric air entering one side of the structure, flowing through the inlet channel between the absorber slabs, flowing through the absorber slabs, and then flowing through the outlet channels to exit from the opposite side of the structure, as illustrated in Figure A-6.

Depending on a number of factors – location, wind speed and direction, available power, etc. – the CO_2 extraction facility size and number of absorber slabs it contains will vary.

FIGURE A-6

DIMENSIONS OF ABSORBER SLAB ASSEMBLY USING 30 CENTIMETER ABSORBER SLABS

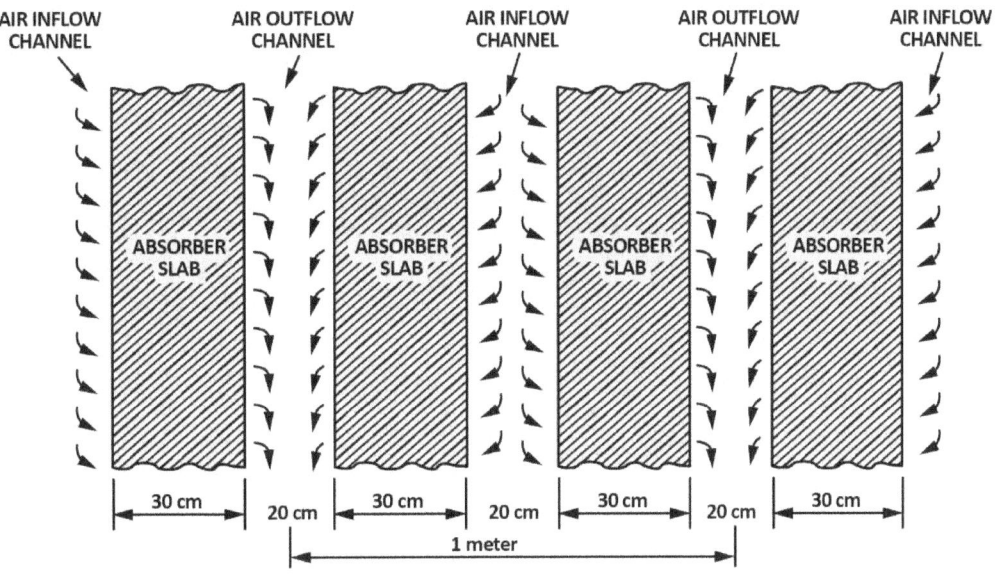

Figure A-7 shows a representative drawing of an absorber slab, with 0.3 meter thickness, H meter height, and L meter length. Typical, height (H) would be on the order of 10 meters. Length L is determined by the acceptable pressure drop for air flowing in the inlet and outlet airflow channels. The maximum air velocity in the inlet air flow channel occurs at the inlet face of the CO_2 extraction facility, while the maximum air flow velocity in the outlet flow channel occurs at the outlet face of the CO_2 extraction facility.

Per meter of height of the absorber slab, the flow areas of the inlet and outlet airflow channels is:

A channel = width of channel x 1 meter height

= 0.2 x 1 = 0.2 m^2

FIGURE A-7
ABSORBER SLAB GEOMETRY

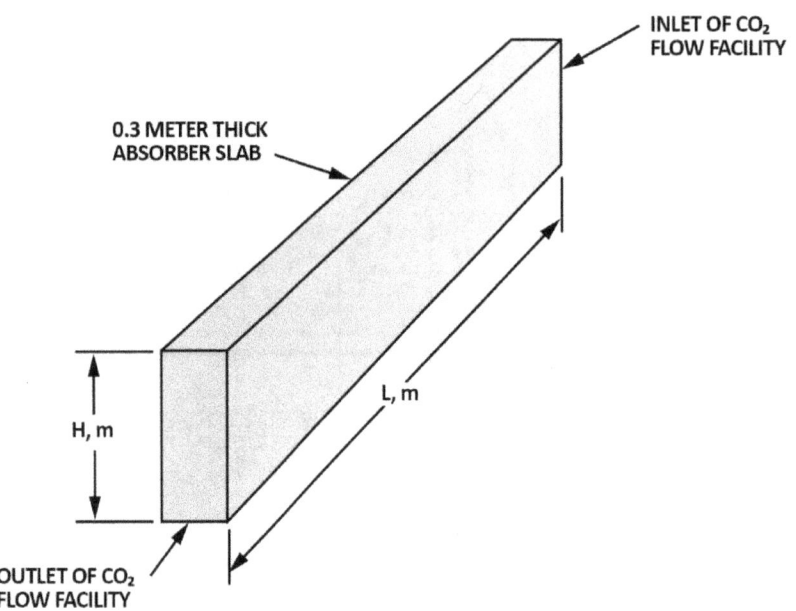

Per meter of length of the absorber slabs with 1 meter of height, the air volume flow rate into the 2 slabs on the sides of the 0.2 m2 channel is:

V_{air} = 2 x inflow velocity x 1m² = 2 x 0.04m/sec x 1m²

= 0.08 cubic meters per second per meter of length

For an absorber slab length of L meters

V_{air} = 0.08 L m/sec

With the maximum flow velocity at the entrance of the CO_2 facility and the maximum flow velocity at its exit equal to

{V_{air}}max = 0.08L/Area of flow channel

= 0.08L/0.2 = 0.4L meters/sec

Flow velocity is linear along the inlet and outlet flow channels, being maximum at the CO_2 entrance point to the inlet flow channel, and zero at its end. For the outlet channel, flow velocity is zero at the CO_2 entrance point and maximum at the CO_2 exit point.

At an intermediate point x in the flow channel, with x measured from the channel entrance

$$\frac{dp}{dx} = \frac{f pair\ V_x^2}{D_o}$$

With

$V_x = V_{max}\left(\frac{x}{L}\right) = 0.4L\left(\frac{x}{L}\right) = 0.4(x)$ meters/sec

At

x = 50 meters, V_{air} = 20 meters/sec, while at x = 25 meters, V_{air} = 10 meters/sec

Integrating, the total Δp for air flow in the channel is

$$\Delta p_{tot} = \frac{f p_{air} (0.4)^2}{D_o} \int_0^L x^2 dx = f_{pair} (0.4)^2 L^3/3D_o$$

Where f = 4x fanning friction factor = approximately 0.01

p_{air} = 1.23 kg/m³

D_o = 2 x 0.2 = 0.4 meter (parallel flow channel 0.2 m wide)

$\Delta p_{tot} = \frac{0.01 \times 1.23}{0.4} (0.4)^2 L^3/3 = 0.00164 L^3$ Pascals

As a function of L, maximum flow velocity x ΔPascals

L(m)	(Vair) max, (m/sec)	ΔP(Pascals)
10	4	1.6
20	8	12.8
30	12	43.2
40	16	102.4
50	20	200
60	24	346
70	28	549

An important conclusion from the above analysis is that the pressure drop in the airflow channels is much smaller than the pressure drop for air flow through the absorber slabs.

For an absorber slab length of 50 meters, for example, the ΔP along the inlet and outlet air flow channels is 200 Pascals, smaller than the 720 Pascals ΔP through the absorber bed.

The air flow rate into the entrance of the CO_2 extraction facility depends on the length L of the absorber slabs.

For an absorber slab length of 50 meters, the air flow rate into the CO_2 facility per square meter of frontal area is:

(Vair) FRONTAL = 0.08L = 4 m3/sec.

The frontal velocity is 4 meters per second, increasing to 20 meters per second as the air inflow begins to flow in the 0.2 meter wide inlet airflow channel. The corresponding Bernoulli ΔP is

$$\Delta P_{Bernoulli} = \frac{1}{2}\rho v^2 = 280 \text{ Pascals}$$

The CO_2 extraction facilities would be located in areas with strong winds, e.g. 5 to 10 meters per second or more to minimize air pumping power.

To illustrate the volume and dimensions of CO_2 extraction facilities, assume that 10,000 facilities are constructed that extract 30 Billion tonnes of CO_2 annually from the atmosphere, about the same amount of CO_2 that is current released to the atmosphere by the combustion of fossil fuels.

Each facility extracts 3 million tonnes of CO_2 per year. Table A-3 lists the design parameters for the facility, while Figure A-8 illustrates its shape and overall dimensions.

The dimensions and area of the CO_2 Extraction Facility (CDEF) are small compared to already exiting facilities. It is only 12 meters high and 54 meters long with a total width of 3.5 kilometers. Total internal volume is 2.2 million cubic meters, approximately 1/6th of the 13.3 million cubic meter volume of the Boeing factory in Everett Washington.

Figure A-9 compares the cross section of the 12 meter x 54 meter physical CO_2 Extraction Facility with the dimensions of existing 3 MW(e) capacity wind turbines. The 3 MW(e) capacity wind turbine has a rotor diameter of 107 meters and a hub height of 100 meter, for a total height above ground of approximately 150 meters, 12 times higher than the 12 meter height of the CO_2 Extraction Facility.

The height of the Statue of Liberty from her foot on the pedestal to the top of her torch, 93 meters, 60 meters less than the top of the 3 MW(e) windmill.

Because the wind only blows about 1/3rd of the time, a 3 MW(e) capacity turbine only puts out an average about 1 MW(e) of power.

If powered by 3 MW(e) capacity wind turbines, the 100 MW(e) CO_2 Extraction Facility would require 100 wind turbines. On average wind turbine farms require about 100 acres of land per 3 MW(e) capacity turbine due to the need for wind turbines not to interfere with each other, access capability, etc. 100 turbines would require a land footprint of 10,000 acres, about 40 square kilometers, 40 million square meters.

The land footprint for the CO_2 extraction facility, 54 meters long and 3.5 kilometers width. Total footprint is only 175,000 square meters. The wind farm land footprint is 40,000,000 square meters. The wind farm land footprint is more than 200 times greater.

Using beamed power from space, described in Chapter 4, the land footprint of the beamed power footprint for 100 MW(e) will be much smaller on the order of 100,000 square meters, comparable to the land footprint of the CO₂ Extraction Facility.

Figure A-10 illustrates a cross sectional view of the absorber slab assembly inside the CO₂ extraction facility. Flexible porous nets attached to the concrete floor of the facility structure are strung to and supported by cables 10 meters above the floor above the floor that run the 54 meter length of the facility.

FIGURE A-8
SHAPE AND OVERALL DIMENSIONS OF 3 MILLION TONNE PER YEAR CO₂ EXTRACTION FACILITY

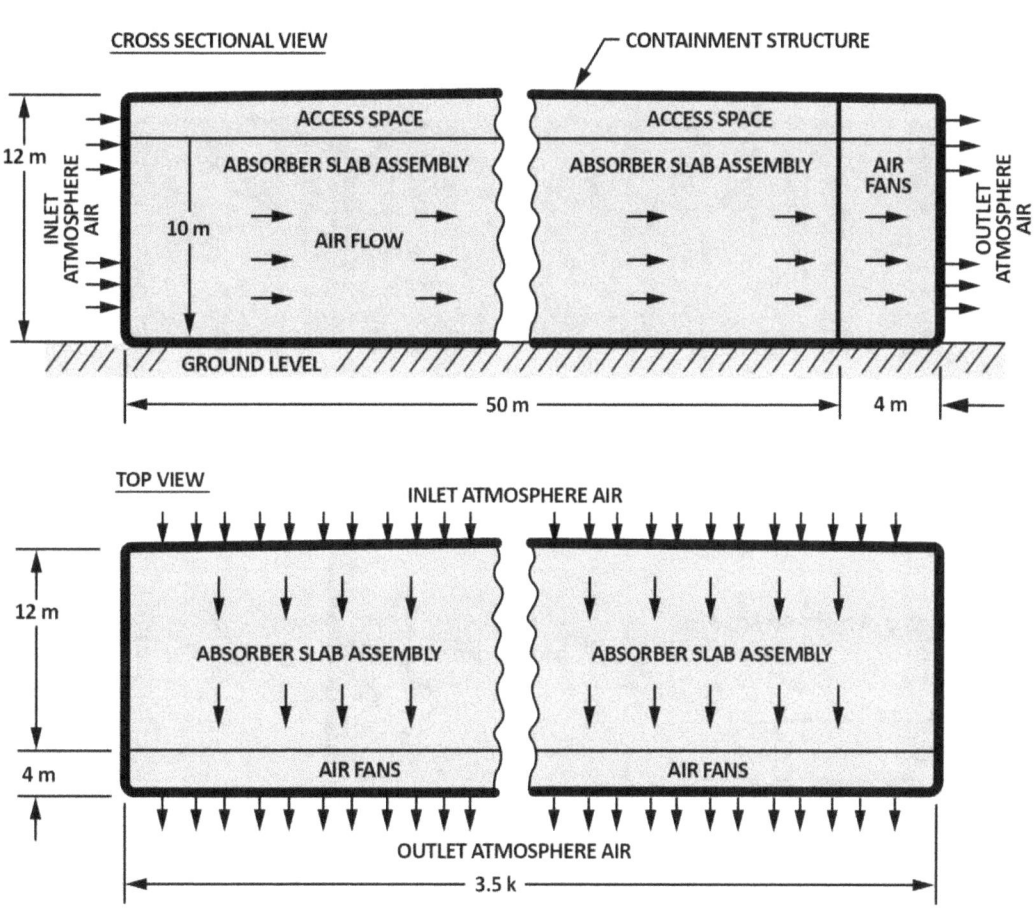

The flexible nets are tensioned to be vertical, and have periodic 2 meter long tensile fibers between their 2 sides to form a 0.2 meter wide, 10 meter high, 54 meter long air flow channel. The nets are closed at their 10 meter top to maintain flow in the channel. The air inflow channel is open at the air entrance to the facility but closed at the end of the channel, at the exit from the facility.

Similarly, the ends of the air outflow channels are closed at the air entrance to the facility and open at the air exit from the facility.

The flexible nets that form the air inflow and outflow channels are porous, so that air can flow transversely outwards from the air inlet flow channels into the absorber slabs, and transversely from the absorber slabs inward to the air outlet flow channels. The pore size is smaller the PEI particle size, so that the particles remain contained in the absorber slab.

The vertical tension on the porous nets, combined with the periodic 0.2 meter long tensile fibers that tie the 2 nets together for a 0.2 meter wide, 10 meter high, 54 meter long air flow channel. Inflatable balloons may also be used at certain points to ensure full channel width.

FIGURE A-9
CO_2 EXTRACTION FACILITY COMPARED TO A 3 MW WINDMILL

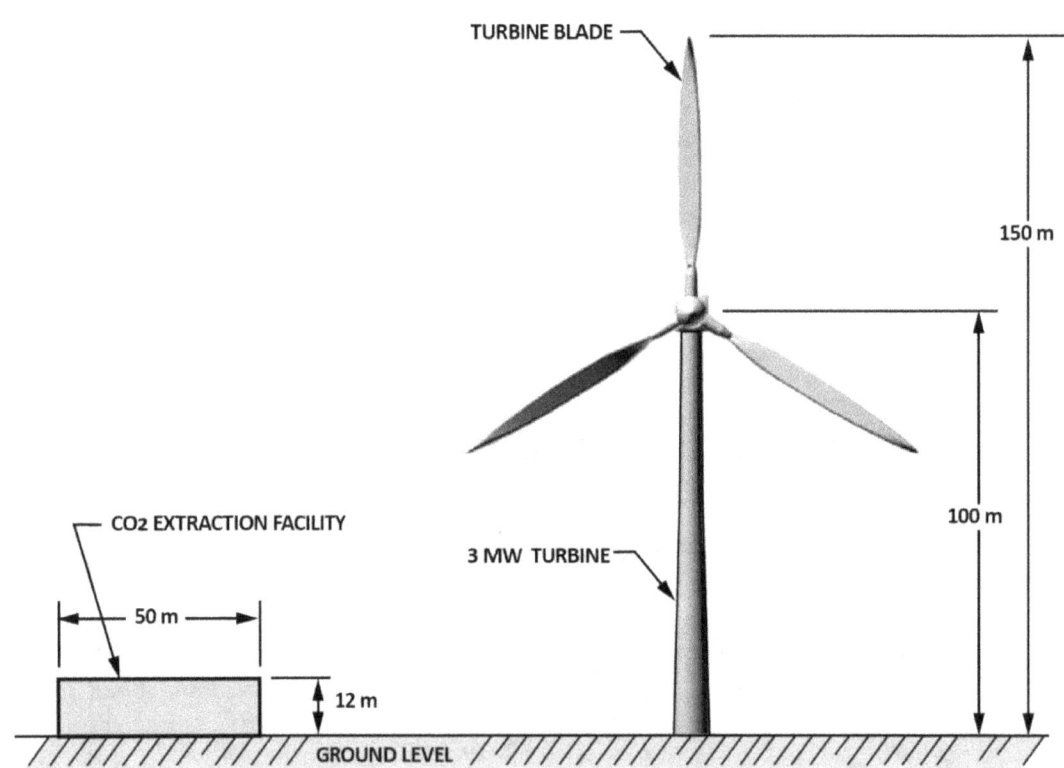

TABLE A-3
DESIGN PARAMETERS FOR A 3 MILLION TONNE PER YEAR CO_2 EXTRACTION FACILITY

- 3 Million Tonnes Per Year CO_2 Extraction From Atmosphere
- 10,000 Facilities For 30 Billion Tonnes Per Year CO_2 Extraction Rate
- 400 ppm CO_2 Concentration In Atmospheric Air Entering CO_2 Extraction Facility
- 350 ppm CO_2 Concentration In Atmospheric Air Leaving CO_2 Extraction Facility
- 1.4×10^5 M^3/Sec Air Inflow Rate Facility
- CO_2 Absorbed By Flow Through Settled Bed Of PEI Particles
- PEI Particles in Absorber Slab With Air Inlet Flow Into 1 Side Of Slab Through It, With Air Outlet Flow On Other Side.
- Absorber Slab Dimensions 0.3 M Thick, 10 m High, 50 m Long
- 0.2 Meter Spacing Between Absorber Slabs For Inlet Air Outflow Channels.
- 0.04 Meter/Sec Superficial Airflow Velocity Through Absorber Slab
- 7000 Absorber Slabs In CO_2 Extraction Facility
- 100 Tonnes Of PEI Absorber Particles Per Absorber Slab
- 5% By Weight CO_2 Absorption At Saturation Of PEI Particle
- 7 Days Cycle Time For PEI Absorption/Desorption Cycle
- 750 Pascals Air Pressure Drop Through Absorber Slab Assembly
- 100 MW(E) Pumping Power For Air Flow Through Facility
- 0.7 Million Tonnes Of PEI Particles In CO_2 Facility
- Overall Dimensions Of CO_2 Extraction Facility
- 12 Meters Height Above Ground
- 54 Meters Long In Air Flow Direction
- 3.5 Kilometers Wide
- 4 Meter Per Sec Air Velocity Into CO_2 Extraction Facility

FIGURE A-10
FLEXIBLE NET ASSEMBLY IN CO₂ EXTRACTION FACILITY

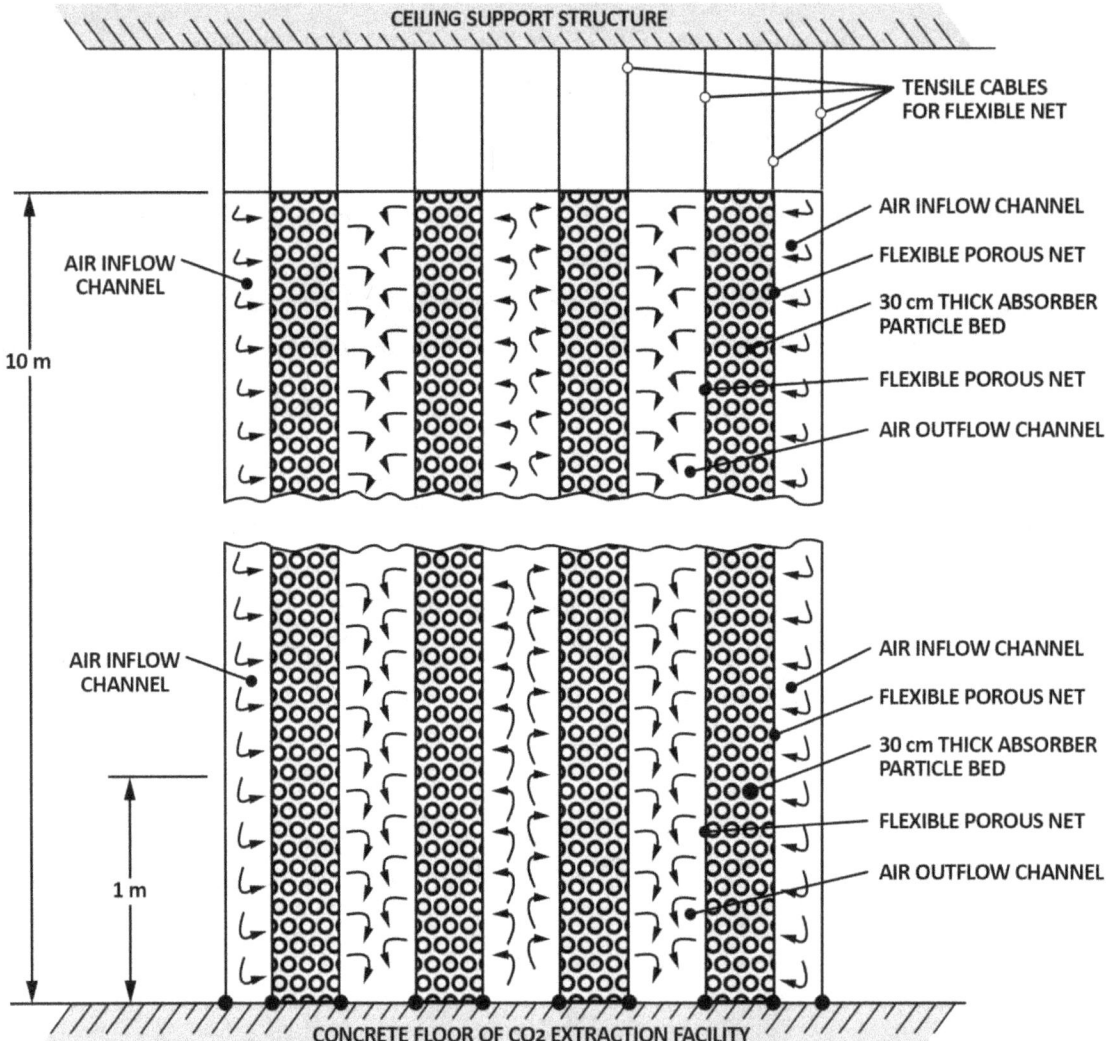

To form the absorber slabs, PEI particles would be loaded into the spaces between the vertical nets that form the inlet and outlet airflow channels. The loading would be done using mobile vehicles operating in the 2-meter-high access space above the 10-meter-high absorber slab assembly.(Figure A-8)

The mobile vehicles would travel on the network of beams positioned above the absorber slab assembly. The beams would also hold the cables to which the vertical porous nets that form the air inlet and outlet flow channels were attached.

The total weight of PEI particles in the CO_2 facility is 0.7 million tonnes, about one fourth of the 2.8 million tonnes of coal burned annually in a 1000 MW(e) power plant.

Operating over a load period of 1 year would require loading 2000 tonnes of PEI particles per day with a load/unload cycle time of 2 hours and 10 tonnes of PEI per vehicle, only 20 mobile vehicles would be needed to do the job.

If necessary, absorber slabs with poor performing PEI particles would be unloaded, using air suction devices.

Detailed engineering analysis, and prototypes will be required to determine the actual capital and operating costs for CO_2 extraction facilities based on the absorber slab approach described here. However, it is possible to project costs based on reasonable assumptions.

We first look at what humanity could afford to spend to extract CO_2 from the atmosphere to prevent runaway global warming and mitigate the severity of the climate changes already occurring due to our use of fossil fuel.

Spending 1 trillion dollars per year to extract 30 Billion tonnes of CO_2 from the atmosphere appears very reasonable. Current world GDP is 78 Trillion dollars annually, and is projected to grow to over 200 Trillion dollars annually by 2050.

Total world annual military expenditures were 1.7 Trillion dollars in 2015 (5), 70% greater than the assumed 1 Trillion dollars for CO_2 removal. In 2015 total world production of motor vehicles was 89 million vehicles (6) at $10,000 average cost per vehicle, annual world spending on vehicles is on the order of 9 Trillion dollars, much greater than the assumed 1 Trillion dollars for CO_2 extraction.

The 1 Trillion dollar annual cost for extraction of 30 Billion tonnes of CO_2 per year corresponds to a cost of $33 per tonne of CO_2. For 10,000 CO_2 Extraction Facilities, the annual cost for each facility would be 100 million dollars including both its amortized capital cost plus operating cost.

The principal capital cost components for the CO_2 extraction facility are:

- Building Structure
- Flexible nets
- Installation of flexible nets to form absorber slabs
- PEI absorber particles
- Air Flow Fans

The principal operating cost components for the CO_2 absorption facility are:

- Power for airflow fans
- Energy for desorption of CO_2 from PEI particles
- Personnel & maintenance expenditures.

To put capital and operating costs in context, suppose that the operating cost was zero. What would total initial capital cost correspond to an annual amortized capital cost of 100 million dollars per year, for the 3 million tonnes of CO_2 per year extraction facility., amortized over a year period?

Initial Capital Cost = 30 years x 100 million dollars/year = 3 Billion Dollars

We first start with the CO_2 extraction building structure. The cost of construction of large industrial building in the US is in the range of 50 to 100 dollars per square foot of floor area depending on size, etc. The floor area of the CO_2 extraction facility is:

AF= 3.5×10^3 m x 50 m = $1.75 \times 10^5 m^2$ = $1.9 \times 10^6 ft^2$

Taking an average cost of $75 per square foot, the CO_2 building structure capital cost would be:

$C_{Building}$ = \$75/ft^2 x 1.9×10^6 = 140 million dollars

Or, 4.7% of 3 Billion dollars

The second capital cost component is the cost of the flexible nets. Each net has an area of

A Net = 10 meter high x 50 meter long = 500 m^2

One meter length of the 3.5 kilometer long CO_2 Extraction Facility has 2 nets, 10 meter x 50 meter in dimensions. The total number of nets is then

Nnets total = $4 \times 3.5 \times 10^3$ = 1.4×10^4

The total surface area of the 14,000 nets is

$A_{surface}$ = 500 x 1.4×10^4 = $7 \times 10^6 m^2$

The flexible net package consists of 2 nets attached together:

A network of high tensile strength plastic fibers that enable it to be pulled up vertically by tensile forces and attached to the oblong structure in the CO_2 building.

An attached porous film with small holes that contain PEI particles but allow air to flow through the film.

Commercial nets cost on the order of $0.1 to $0.2 per square meter (8). Even if the CO_2 extraction nets were to cost 10 times as much, e.g., $2 per square meter, the capital cost of the 7×10^6 m^2 of nets would be only 14 million dollars only 0.5% of 3 Billion dollars. Even if it were to be 10 dollars per square meter, which appears far too much, their capital cost of 70 million dollars would be very small compared to 3 Billion dollars.

For this study, $2 per m^2 for the net cost appears reasonable, and is the value adapted for capital cost estimation.

The 3rd capital cost component is the installation cost of the flexible nets. The installation involves the following steps.

Layout of the flexible net package (0.2 meter wide, 10 meter high, 50 meter long) along the floor of the 3.5 kilometer long, 50 meter wide CO_2 extraction facility.

Attachment of the bottom of flexible net package to the floor

10-meter vertical pullup of the top of the flexible net package and attachment to beams at the top of the CO_2 building structure.

The capital cost for installing the 7,000 flexible net packages in the facility depends on the manhours required and the cost per man-hour. The worker will install the flexible nets using vehicles that carry them plus conventional cranes. Assuming a 6 month erection period, on the order of 35 nets would be erected daily.

Assuming an 8 hour period to erect a flexible net, 4 erection teams would be required. Based on 8 workers per team, that is 32 workers and 250 manhours per day, 46,000 man-hours over the 6 month period.

At $50 per man hour, the total flexible net installation cost would be 2.3 million dollars, a very small amount compared to other capital cost components. Amortized over a 30 year period it is only about $10,000 per year. For a 3 million tonne per year CO_2 extraction rate from the CO_2 facility it amounts to only $470,000/3 \times 10^6$ = 2 cents per tonne of CO_2

The 4th capital cost component is the cost of the PEI particles. The CO2 Extraction Facility design described here (Table A-3) requires 0.7 million tonnes of PEI particles. Currently, the PEI particles are only fabricated in very small amounts for experimental testing, so that it is difficult to project what their cost will be if produced in very large quantities – billons of tonnes.

Assuming that the PEI particles will cost on the order of most plastics, $1000 per tonne ($1 per Kg), their cost for the CO_2 extraction facility would be 0.7×10^6 tonnes x $1000/tonne, equal to 0.7 Billion dollars, the largest capital cost component.

The 5th capital cost component is the cost of loading the PEI particles into the 7,000 absorber slabs. As described earlier, the PEI particles are loaded into the spaces between the installed flexible net packages to form the particle be absorber slabs.

The PEI particles are loaded into the absorber slab spaces by mobile vehicles traveling on the beams above the installed flexible net packages, using air blowers that direct the PEI particles into the absorber slab spaces.

As described earlier, loading 0.7 million tonnes of PEI particles into the absorber slabs over a period of 1 year corresponds to 2000 tonnes per day. 20 mobile vehicles operating with 10 tonnes of particles per vehicle and a 2 hour load/unload cycle could load 2000 tonnes per day With 20 vehicles and a shift force of 100 persons for operation, with 3 shifts daily, total man hours per day would be 3 x 100 = 300 manhours daily. Over 1 year total manhours would be 110,000. At $50 per manhour the capital cost for loading the PEI particles would be 5.5 million dollars, a small fraction of total capital cost.

The 6th principal capital cost component is the cost of the industrial fans that move the air through the CO_2 facility. Total air pump power is 100 MW(e)(Table A3), with an airflow rate of 1.4×10^5 cubic meters per second.

Current large industrial fans have airflow rates of up to about 1,000 cubic meters per second.(9) For the CO_2 facility this would require on the order of 100 fans, spaced along its 3.5 kilometer length, one every 35 meters.

There is very little available data in the capital cost of large industrial fans. As a surrogate, we use the capital cost of gas turbines. The capital cost of gas turbines is typically about $200 per kilowatt.(10)

For 100 MW(e) of fan power at $200/kw, the capital cost would be 10^5 kw(e) x 200/kw(e) = 100 million dollars, an acceptable amount.

Table A-4 summarized the principal capital cost components for the 3 million tonnes per year CO_2 extraction facility. The amortized capital cost per year is 32 million dollars per year, corresponding to approximately $10 per tonnes of CO_2.

The amortized capital cost is very attractive. It could be substantially greater and still very acceptable. The only capital cost component for which changeds in its projected capital cost could be significantly alter the total cost is the cost of the PEI particles. The other components are much smaller, and more defined in price.

We now turn to the operating costs for the CO_2 Extraction Facility.

The first operating cost component is the cost of the pumping power for the airflow fans. At current industrial power costs of about 6 cents per kwhr, 100 MW(e) of fan power would cost 52 million dollars per year, a very acceptable amount. With beamed power from space solar power satellites at 2 cents per kwhr(e), as described in Chapter 4, the fan power cost would only cost 17 million dollars per year, a factor of 3 cheaper.

The second operating cost component is the cost of the energy needed to periodically heat the PEI particles to the temperature required to desorb the CO_2 that they have absorbed from the atmosphere and flowing through the absorber slab.

TABLE A-4
SUMMARY OF CAPITAL COST COMPONENTS FOR 3 MILLION TONNE PER YEAR CO_2 EXTRACTION FACILITY

Capital Cost Component	Amount, millions of Dollars
CO_2 Facility Building Structure	140
Flexible Nets	3
Installation of Flexible Nets	2.3
PEI Absorber Particles	700
Loading of PEI absorber Particles Into Absorber Slabs	5.5
Air Flow Fans	100
	Total: 950 million Dollars
Amortized Cost Over 30 years per operating year	32 million dollars/year
Amortized Capital Cost Per Tonne of CO_2 Extracted	$10/tonne of CO_2

There are 0.7 million tonnes of PEI particles in the CO_2 extraction facility. During CO_2 absorption, the particles operate at a temperature of about 20 Degrees Centigrade. To desorb the CO_2 the then require heating to about 80 Degrees Centigrade.

Per tonne of PEI particles, with a heat capacity of ~ 1 J/g°C, the energy required is:

ΔH_{th} = 10^6 grams x 1 J/g°C x 80°C

For 0.7 million tonnes

ΔH_{th} = 17 x 0.7 x 10^6 = 1.2 x 10^7 kwhr(th)

For 52 cycles per year for the CO_2 absorption/desorption process

ΔH_{th} = 1.2 x 10^7 x 52 = 7.61 x 10^8 kwhr(th)

The total electric energy for the 100 MW(e) (10^5 kw(e) powered air flow fans is

ΔE = 10^5 kw(e) x 8760 hours = 8.8 x 10^8 kwh (e)

The thermal energy for heating the PEI particles could be supplied electrically, solar heaters, or heat pumps. Priced at the same cost per kwhr as electric power, the cost of the thermal energy to desorb CO_2 would be slightly less than the electric power for the airflow plans – on the order of 15 million dollars per year at 2 cents per kwhr(th).

The 3rd operating cost component is personnel for operations maintenance and equipment replacement and maintenance.

For personnel, an operating staff of 150 people appears sufficient for 3 shift operation. At $100,000 per person per year including benefits, the annual personnel costs would be 15 million dollars.

The cost of replacement equipment should be small, with the principal components being maintenance of the airflow fans and replacement of failed PEI particles. Without extended testing of PEI absorber slab prototypes, it is difficult to know if they will need periodic replacement, and if so, how often. Annual maintenance cost for the air fow fans is taken as 10% of their capital cost, that is 10 million dollars per year.

For example, a 10-year lifetime for the PEI particles would require an annual cost at $1000 per tonne is

Annual Cost PEI Replacement: = 0.7 x 10^6 tonnes x $1000/tonne/10 years

= 70 million dollars

With a corresponding cost of

Cost/tonne of CO_2 = 70 million$/3 million tonnes of CO_2

= $23/tonne of CO_2

Still acceptable, but it points out the need for extended testing of CO_2 extraction prototype

Assuming that the PEI particles last for 30 years, the total operating cost is given in Table A-5, as 57 million dollars per year.

TABLE A-5

SUMMARY OF OPERATING COST COMPONENTS FOR 3 MILLION TONNE/YEAR CO_2 EXTRACTION FACILITY

Operating Cost Component	Annual Amount, Millions $/year
Electric Power for Fans @ 2 cents/kwhr(e)	17
Thermal Energy for Heating PEI particles@ 2 Cents/kwhr(e)	15
Operating Personnel and Maintenance	25
	Total = 57 million$/year
Operating Cost per tonne of CO_2 Extracted	$19 per tonne CO_2
Total Cost per tonne, Amortized Capital + operating	$29/tonne of CO_2

Adding in the amortized capital cost of 32 million dollars per year from Table A-4, the total annual cost for the 3 million tonne per year CO_2 extraction facility is

Total Annual Cost = 32 million $ + 57 million $ = 89 million

Equal to $29 per tonne of CO_2

For the global assembly of 10,000 CO_2 extraction facilities that extract 30 Billion tonnes of CO_2, annually. Table A-6 shows the total would cost per year of 860 Billion dollars per year, $29 per tonne of CO_2 extracted.

860 Billion$ per year. A lot of money but just 1/2 of the World military expenditures. About 1% of present World GDP, and 0.3% of projected World GDP in 2050. Clearly affordable.

To not reduce CO_2 concentration in the atmosphere when it is affordable would be insane.

TABLE A-6

TOTAL COST FOR EXTRACTION OF 30 BILION TONNES PER YEAR FROM THE ATMOSPHERE USING 10,000 EXTRACTION FACILITIES

Capital Cost Component	Capital Cost amount, Trillion
CO_2 Buildings	1.4
Flexible Nets	0.03
Installation of Flexible Nets	0.02
PEI Absorber Particles	7.0
Loading of PEI Absorber Particles	0.6
Air Flow Fans	1.0
Total Capital Cost	9.5 Trillion $
Amortized Annual Cost (30 year Amortization)	320 Billion$/year
Operating Cost Component	Annual Cost, Billion$/Year
Electric Power for Air Flow Fans	170
Thermal Energy for CO_2 Desorption	150
Operating Personnel and Maintenance	250
Total	570 Billion$/year
Total Annual Cost (Amortized Capital + Operating)	890 Billion $/year
Cost/Tonne of CO_2 Extracted	$29/tonne of CO_2

List of References

1. www.gwcc.net/global-figures/wind-in-numbers

2. Capturing CO_2 from Ambient Air Using A Polyethyleneimine – Silica Absorbent in Fluidized Beds, Zhang, et al, Chemical Engineering Science, 116, 6 September, pages 306-316

3. https://en.wikipedia.org/wiki/List_of_largest_buildings

4. https://en.wikipedia.org/wiki/list_of_countries_by_number_of_households

5. http://en.wikipedia.org/wiki/list_of_countries_by_military_expenditures

6. https://en.wikipedia.org/wiki//automotive_industry

7. The DCD Industrial Buildings Cost Per Square :Foot Analysis
http://www.dcd.com/pdf_files/1107analysis.pdf

8. https://www.alibaba.com/showroom/mesh-netting-roll/html/

9. https://en.wikipedia.org/wiki/Industrial_fan

10. http://www.gas-turbines.com/trader/kwprice.htm

List of Figure Credits

A-1. Process for Absorbing Carbon Dioxide From The Atmosphere Using Packed Beds of Absorber Particles, Author, Powell

A-2. Flow Path of Atmospheric Air Through Packed Particle Bed, Author Powell

A-3. Arrangement of Multiple Absorber Slabs With Inflow and Outlfow Air Channels, Author, Powell

A-4. Absorber Bed Geometry for Optimization of Operating Parameters, Author, Powell

A-5. Absorber Thickness and Pressure Drop Across It As A Function of Superficial Flow Velocity, Author, Powell

A-6. Dimensions of Absorber Slab Assembly Using 30 Centimeter Absorber Slabs, Author, Powell

A-7. Absorber Slab Geometry, Author, Powell

A-8. Shape and Overall Dimensions of 3 Million Tonne Per Year CO_2 Extraction Facility, Author, Powell

A-9. Comparisons of Dimensions of CO_2 Extraction Facility With Existing Wind Turbine, Author, Powell for CO_2 Facility

A-10. Absorber Slab Assembly Inside the CO_2 Extraction Facility

Appendix B

Bibliography of Papers and Reports

on:

StarTram (Maglev Launch)

MITEE (Nuclear Thermal Propulsion Engine)

SUSEE (Nuclear Space Power Reactor)

ALPH (Nuclear Robotic Probe and Factory)

MIC (Magnetically Inflated Cable Space Structures)

Detailed Papers and Reports on StarTram

StarTram: A New Concept for Very Low-Cost Earth-to-Orbit Transport Using Ultra High Velocity Magnetic Launch, James Powell, George Maise, and John Paniagua, Paper IAF-01-S.6.04, 52nd International Astronautical Congress, 1-5 October 2001, Toulouse, France (36 pages).

Powell, J., and Maise, G., "Space Tram" US Patent No. 6,311,926B1, November 6, 2001.

StarTram: A New Approach for Low-Cost Earth-to-Orbit Transport, James Powell, George Maise, and John Paniagua, 2002 IEEE Space Conference, March 2002, Big Sky, Montana (17 pages).

StarTram C—A Maglev System for Ultra Low Cost Launch of Cargo to LEO, GEO, and the Moon, James Powell, George Maise, and John Paniagua, Paper IAC-03-IAA13.1.04, 54th International Astronautical Congress, October 2003, Bremen, Germany (18 pages).

StarTram: The Key to a Robust, Low Cost Earth/Lunar Transport System, James Powell, George Maise, and John Paniagua, International Lunar Conference 2003, November 16-22, 2003, Hawaii (23 pages).

StarTram: Ultra Low-Cost Launch for Large Space Architectures, James Powell, George Maise, and John Paniagua, STAIF 2004 Conference, February 2004, Albuquerque, New Mexico (12 pages).

StarTram: The Key to Low Cost Lunar Bases and Human Exploration of Space, James Powell, George Maise, and John Paniagua, AIAA Space 2004, September 28-30, 2004, San Diego, California (12 pages).

StarTram: An Ultra Low Cost Launch System for Large Scale Exploration and Commercialization of Space, James Powell, George Maise, and John Paniagua, Paper IAC-04-V.05.07, 55th International Astronautical Congress, October 2-8, 2004, Vancouver, Canada (17 pages).

Ibid, StarTram viewgraphs presented at the 55th IAC meeting, Vancouver, Canada, (26 pages).

StarTram: An Ultra Low Cost Launch System to Enable Large Scale Exploration of the Solar System, James Powell, George Maise, and John Paniagua, STAIF 2006 Conference, February 12-15, 2006, Albuquerque, New Mexico (12 pages).

StarTram: An International Facility to Magnetically Launch Payloads at Ultra Low Unit Cost, George Maise, James Powell, John Paniagua, and James Jordan, Paper IAC-06-D3.2.7, 57th International Astronautical Congress, October 2-5, 2006, Valencia, Spain (14 pages).

Ibid: StarTram viewgraphs presented at the 57th IAC Meeting, Valencia, Spain (25 pages).

StarTram: The Maglev Launch Path to Very Low Cost, Very High Volume Launch to Space; presented at the 14th International EML Symposium, Victoria, Canada, June 10-13, 2008.

The Gen-1 Maglev Launch System for Ultra Low Cost Access to Space; James Powell, George Maise, and John Paniagua; presented at the 59th International Astronautical Congress (IAC), Glasgow, Scotland, September 29—October 3, 2008 (10 pages).

Ibid: Viewgraphs presented at the 59th IAC Meeting, Glasgow, Scotland (21 pages).

Maglev Launch—An Ultra Low Cost Way to Deploy Space Solar Power Systems; presented at the From the Sun to the Earth International Conference on solar Energy from Space, Ontario Science Center, Toronto, Canada, September 8-10, 2009 (17 pages).

Ibid: Viewgraphs presented at the From the Sun to the Earth Conference, Toronto, Canada (33 pages).

Maglev Launch: Ultra Low Cost, Ultra/High Volume Access to Space for Cargo and Humans; James Powell, George Maise, and John Rather, presented at SPESIF-2010—Space, Propulsion, and Energy Sciences International Forum, February 23-26, 2010, Johns Hopkins Applied Physics Laboratory, Baltimore, Maryland (15 pages).

Ibid: viewgraphs presented at SPESIF-2010 meeting, Baltimore, Maryland (33 pages).

A Development and Test Program for the Generation-1 Maglev Launch System, James Powell, George Maise, and John Rather, presented at SPESIF-2011 Meeting, Baltimore Maryland (16 pages).

Ibid: Viewgraphs presented at SPESIF-2011 Meeting, Baltimore, Maryland.

Detailed Papers and Reports on MITEE

MITEE: An Ultra Lightweight Nuclear Engine for New and Unique Planetary Science and Exploration Missions. James Powell, John Paniagua, George Maise, Hans Ludewig and Michael Todosow, Paper IAF-98-R.1.01, 49th International Astronautical Congress, Sept. 28—Oct. 2, 1998, Melbourne, Australia [27 pages].

Europa Sample Return Mission Utilizing MITEE Technologies. John Paniagua, James Powell, George Maise, Hans Ludewig, Michael Todosow, Paper IAF-98-Q.2.03, 49th International Astronautical Congress, Sept. 28—Oct. 2, 1998, Melbourne, Australia [15 pages].

Exploration of Jovian Atmosphere Using Nuclear Ramjet Flyer. George Maise, James Powell, John Paniagua, Hans Ludewig, Michael Todosow, Paper IAF-98-S.6.08, 49th International Astronautical Congress, Sept. 28—Oct. 2, 1998, Melbourne, Australia [11 pages].

High Performance Nuclear Thermal Propulsion System for Near Term Exploration Missions to 100 AU and Beyond. James Powell, John Paniagua, George Maise, Hans Ludewig, and Michael Todosow, Acta Astronautica, 44 No. 2-4, pp 159-166, Jan—Feb. 1999 [8 pages].

The Liquid Annular Reactor System (LARS) for Deep Space Exploration. George Maise, John Paniagua, James Powell, Hans Ludewig, and Michael Todosow, 2nd IAA Symposium on Realistic Near-Term Advanced Scientific Space Missions. June 29—July 1, 1998, Aosta, Italy; also Acta Astronautica, 44 No. 2-4, pp 167-174, Jan—Feb 1999 [13 pages].

New Approaches for the Exploration and Colonization of the Solar System: Road Map for the Next 30 Years in Space. James Powell, George Maise, and John Paniagua, Report PUR-7, Nov. 10, 1998 [18 pages].

The MITEE Family of Compact, Ultra Lightweight Nuclear Thermal Propulsion Engines for Planetary Space Exploration. James Powell, George Maise, and John Paniagua, Paper IAF 99-5.6.03, 50th International Astronautical Congress, October 4-8, 1999, Amsterdam, the Netherlands [28 pages].

SunBurn: A Concept Enabling Ultra High Spacecraft Velocities for Extra Solar System Exploration. George Maise, James Powell, and John Paniagua, Paper IAA-99-IAA.4.1.07, 50th International Astronautical Congress, October 4-8, 1999, Amsterdam, the Netherlands [18 pages].

A Cost Effective Space Infrastruction for Retrieval of Helium-3 from Uranus for Earth-Based Fusion Power Systems Utilizing the MITEE Nuclear Propulsion System. John Paniagua, James Powell, and George Maise, Paper IAA-99-R.3.10, 50th International Astronautical Congress, October 4-8, 1999, Amsterdam, the Netherlands [17 pages].

Compact, Ultra Lightweight Nuclear Thermal Propulsion Engines for Planetary Science Missions. James Powell, George Maise, John Paniagua, Hans Ludewig, and Michael Todosow, 10th Annual NASA/JPL/MFSC/AIAA Advanced Propulsion Research Workshop, Huntsville, Alabama, April 6-8, 1999 [25 pages].

Phase 1 Final Report: Lightweight High Specific Impulse (1000 sec) Space Propulsion Systems. James Powell, George Maise, John Paniagua, Jon Longtin, John Metzger, and Hui Zhang, PUR-12, October 1999 [221 pages].

Phase 1 NIAC Final Report: Exploration of Jovian Atmosphere Using Nuclear Ramjet Flyer. George Maise, James Powell, John Paniagua, and Robert Lecat, PUR-16, Nov. 30, 2000 [14 pages].

MITEE-B: A Compact Lightweight Bi-Modal Nuclear Engine to Deliver Both High Propulsive Thrust and High Electric Power. James Powell, George Maise, John Paniagua, and Stan Borowski, Paper IAF-01-S.6.05, 52nd International Astronautical Congress, Oct. 1-5, 2001, Toulouse, France [24 pages].

Europa One—A Manned Base for Exploration of the Outer Solar System and Near Interstellar Space. John Paniagua, James Powell, and George Maise, 52nd International Astronautical Congress, October 1-5, 2001, Toulouse, France [26 pages].

Phase 1 NIAC Final Report: Europa Sample Return Mission Utilizing High Specific Impulse Propulsion Refueled with Indigenous Resources. John Paniagua, James Powell and George Maise, November 30, 2001, Report PUR-21 [114 pages].

Compact MITEE-B: Bi-Modal Nuclear Engine for Unique New Planetary Science Missions. James Powell, George Maise, John Paniagua, and Stanley Borowski, AIAA 2002-3652, AIAA/ASME/SAE/ASEE Joint Propulsion Conference and Exhibit, July 2002, Indianapolis, Indiana [22 pages].

Phase 2 NIAC Interim Report: Exploration of Jovian Atmosphere Using Nuclear ramjet Flyer. George Maise, et al, PUR-26, Jan. 31, 2002 [125 pages].

Europa Sample Return Mission Utilizing High Specific Impulse Refueled with Indigenous Resources. John Paniagua, James Powell, and George Maise, Paper IAC-

02-Q.2.05, 53rd International Astronautical Congress, October 10-19, 2002, Houston, Texas [14 pages].

Missions Possible: How Humanity Can Really Explore the Solar System Using Nuclear Propulsion, Report PUR-27, James Powell, George Maise, and John Paniagua, April 15, 2002 [22 pages].

Bi-Modal MITEE Engine for Nuclear Thermal/Nuclear Electric Propulsion. James Powell, George Maise, and John Paniagua, Advanced Space Propulsion Workshop, June 4-6, Pasadena, California [31 pages].

Exploration of Jovian Atmosphere Using Nuclear Ramjet Flyer. James Powell, George Maise, and John Paniagua, Advanced Space Propulsion Workshop, June 4-6, Pasadena, California [31 pages].

NEMO: Exploration of Europa's Subsurface Ocean and Return of Samples to Earth Using Nuclear Propulsion. James Paniagua, James Powell, and George Maise, Advanced Space Propulsion Workshop, Huntsville, Alabama, April 15-17, 2003 [35 pages].

MITEE-B: A Compact Ultra Lightweight Bi-Modal Nuclear Propulsion Engine for Robotic Planetary Science Missions. James Powell, George Maise, John Paniagua, and Stanley Borowski, STAIF 2003 Meeting, February 2003, Albuquerque, New Mexico [9 pages].

Pluto Orbiter/Lander/Sample Return Missions Using the MITEE Nuclear Engine. James Powell, George Maise and John Paniagua, 2003 IEEE Aerospace Conference, Big Sky, Montana, March 2003 [24 pages].

Exploration of Jovian Atmosphere Using Nuclear Ramjet Flyer. George Maise, et al., Phase II Final Report, March 1, 2003, NIAC Phase II Grant 07600-061 [163 pages].

HIP: A Hybrid NTP/NEP Propulsion System for Ultra Fast Robotic Orbiter/Lander Missions to the Outer Solar System. James Powell, George Maise, and John Paniagua, 54th International Astronautical Congress, October 2003, Bremen, Germany [17 pages].

MITEE and SUSEE: Compact Ultra Lightweight Nuclear Power Systems for Robotic and Human Exploration Mission. James Powell, George Maise, and John Paniagua, Paper IAC-04-IAA R.4/S.7-04, 55th International Astronautical Congress, Vancouver, Canada, October 2-8, 2004 [15 pages].

NEMO: A Mission to Explore and Return Samples from Europa's Oceans. James Powell, John Paniagua, and George Maise, STAIF 2004 Conference, February 2004, Albuquerque, New Mexico [7 pages].

Nuclear Propulsion and Power Systems for Near Term Exploration of the Solar System. James Powell, George Maise, and John Paniagua, AIAA 1st Space Exploration Conference, Jan 30—Feb. 1, 2005, Orlando, Florida [17 pages].

NEMO: A Mission to Search for and Return to Earth Possible Life Forms on Europa, Jesse Powell, James Powell, George Maise, and John Paniagua, Acta Astronautica, 57 pp 579-593, 2005 [15 pages].

Mini-MITEE: Ultra Small, Ultra-Light NTP Engines for Robotic Science and Manned Exploration Missions. James Powell, George Maise, and John Paniagua, STAIF 2006 Conference, February 12-16, 2006, Albuquerque, New Mexico [10 pages].

MITEE: A Compact Near Term NTP Engine for New and Unique Robotic and Manned Exploration Missions. James Powell, George Maise, and John Paniagua, American Nuclear Space Conference, June 5-9, 2005, San Diego, California [9 pages].

The MITEE Hopper: A Compact NTP Spacecraft to Explore Multiple Surface Sites Using In-Situ Propellants. James Powell, George Maise, and John Paniagua, Paper IAC-06-D2.8/C3.5/C4.7/D3.5.06, 57th International Astronautical Congress, October 2-6, 2006, Valencia, Spain [12 pages].

A New Mission for the International Space Station (ISS) Enabled by Nuclear Thermal Propulsion—Cyclic Transport of Personnel and Supplies Between the Earth and the Moon, John Paniagua, James Powell, and George Maise, STAIF 2008 Conference, February 2008, Albuquerque, New Mexico [9 pages].

Design and Development of the MITEE-B Bi-modal Nuclear Propulsion Engine, John Paniagua, James R. Powell, and George Maise.

Application of the MITEE Nuclear Ramjet for Ultra Long Range Flyer Missions in the Atmospheres of Jupiter and Other Giant Planets, George Maise, James Powell, John Paniagua, Edward Kush, Pasquale Sforza, and Hans Ludewig.

Detailed Papers and Reports on SUSEE

SUSEE: Ultra-Light Nuclear Space Power Using the Steam Cycle, James Powell, George Maise, and John Paniagua, IEEE 2002 Space Conference, Big Sky, Montana, 2002 [17 pages].

SASSE: A Lightweight, High Efficiency Solar Thermal Steam Cycle For Satellites, James Powell, George Maise, and John Paniagua, Paper IAC-03-R.2.07, 54th International Congress, Bremen, Germany, 2003 [16 pages].

SUSEE—An Ultra Lightweight Nuclear Electric Propulsion System Based on Existing Water and Steam Cycle Technology, James Powell, George Maise, and John

Paniagua, Advanced Space Propulsion Workshop, Huntsville, Alabama, April 15-17, 2003 [31 pages].

Compact Ultra Light Nuclear Electric Power Systems for Future Moon Bases and Colonies, James Powell, George Maise, and John Paniagua, International Lunar Conference 2003, Hawaii, November 16-22, 2003 [13 pages].

MITEE and SUSEE: Compact Ultra Lightweight Nuclear Power Systems for Robotic and Human Exploration Missions, James Powell, George Maise, and John Paniagua, Paper IAC-04-IAA-R.4/S.7-04, 55th International Astronautical Congress, October 2-8, 2004, Vancouver, Canada [15 pages].

Ibid: Viewgraph presentation [29 pages].

SUSEE: An Ultra Light Space Nuclear Power System Based on Conventional Water Reactor Technology, George Maise, James Powell, and John Paniagua, American Nuclear Society, June 5-9, 2005, San Diego, California [9 pages].

SUSEE: A Compact, Lightweight Space Nuclear Power System Using Present Water Reactor Technology, George Maise, James Powell, and John Paniagua, STAIF 2006 Conference, February 12-16, 2006, Albuquerque, New Mexico [12 pages].

Detailed Papers and Reports on ALPH

ALPH—A Robotic Precursor to Produce Large Amounts of Supplies for Manned Outposts on Mars, James Powell, John Paniagua, and George Maise. Presented at 49th International Astronautical Congress, Melbourne, Australia, September 28—October 2, 1998, Paper IAF-98Q.3.08 [38 pages].

MICE: A Compact, Light Near Term Mobile Robot for Exploration of the Martian Polar Ice Cap, James Powell, George Maise, John Paniagua, Hans Ludewig, and Michael Todosow, Presented at 50th International Astronautical Congress, Amsterdam, the Netherlands, October 4-8, 1999, Paper IAF 99-Q.3.08 [18 pages].

Development of Self-Sustaining Mars Colonies Utilizing North Polar Cap and the Martian Atmosphere, James Powell, George Maise, John Paniagua, and Jesse Powell, Final Report, NIAC Research Grant 07600-053, November 20, 2000 [184 pages].

Self-Sustaining Mars Colonies Utilizing the North Polar Cap and Martian Atmosphere, James Powell, George Maise, and John Paniagua, Presented at 51st International Astronautical Congress, Rio de Janeiro, Brazil, Oct. 2-6, 2000, also published in Acta Astronautica, 48, No. 5-12, pp. 737-765 (2001) [27 pages].

The Mars Hopper—A Mobile Lightweight Probe to Explore and Return Samples from Many Widely Separated Locations on Mars, James Powell, George Maise, and John

Paniagua, Presented at 52nd International Astronautical Congress, Toulouse, France, Oct. 1-5, 2001, Paper IAA-01-IAA13.3.08 [26 pages].

Fast Track Route to Mars Colony Using Nuclear Propulsion and Power, James Powell, George Maise, and John Paniagua, Presented at 40th Aerospace Sciences Meeting and Exhibit, Reno, Nevada, Jan. 14-17, 2002, Paper AIAA 2002-0996 [32 pages].

CADMUS—A Robotic Mars Factory Returning Supplies to Earth Orbit, James Powell, George Maise, and John Paniagua, Presented at 2003 IEEE Aerospace Conference, Big Sky, Montana, March 2003 [18 pages].

Xanadu: A Polar Base for Manufacturing Supplies on Mars, James Powell, George Maise, and John Paniagua, Presented at STAIF 2004 Conference, Albuquerque, New Mexico, February 2004 [9 pages].

Multi-MICE: A Network of Interacting Nuclear Cryoprobes to Explore Ice Sheets on Mars and Europa, Jesse Powell, James Powell, George Maise, and John Paniagua, Presented at Space 2005 Conference, Long Beach, California, Sept. 2005 [14 pages].

MERIT: A New Approach for a Large Scale Space Infrastructure Based on Mars, James Powell, George Maise, and John Paniagua, Presented at 2005 STAIF Conference, February 13-17, 2005, Albuquerque, New Mexico [10 pages].

ALPH: A Low Risk, Cost Effective Approach for Establishing Manned Bases and Colonies on Mars, James Powell, George Maise, John Paniagua, and Jesse Powell, Presented at AIAA Space 2005 Conference, Long Beach, California, August 30—Sept. 1, 2005 [17 pages].

ALPH: A compact Robotic Nuclear Powered Factory to Build and Supply Bases on Mars Prior to Manned Landing, James Powell, George Maise, and John Paniagua, Present at American Nuclear Society Space Nuclear Conference, San Diego, California, June 5-9, 2005 [17 pages].

Multi-MICE: A Network of Interactive Nuclear Cryoprobes to Explore Ice Sheets on Mars and Europa, George Maise, James Powell, Jesse Powell, John Paniagua and Hans Ludewig, NASA Institute of Advanced Concepts, Phase 1 Report, NIAC Subaward No. 07605-003-047, May 1, 2006 [145 pages].

Multi-MICE: Nuclear Powered Mobile Probes to Explore Deep Interiors of the Ice Sheet on Mars and the Jovian Moons, George Maise, James Powell, Jesse Powell, John Paniagua, and Hans Ludewig. Presented at STAIF 2007 Conference, Albuquerque, New Mexico, February 11-15, 2007 [10 pages].

MICE: A System of Compact Mobile Nuclear Probes to Explore the Deep Interior of Mars North Polar Cap, George Maise, James Powell, John Paniagua, Jesse Powell, and Hans Ludewig, presented at the 57th International Congress, Valencia, Spain, October 2-5, 2006, paper IAC-06-A3.P3.5.

Detailed Papers and Reports on MIC

MIC—A Self Deploying Magnetically Inflated Cable System for Large Scale Space Structures, James Powell, George Maise, and John Paniagua. Acta Astronautica 48, No 5-12, pp 331-352, 2001 [21 pages].

Deployment of Large Structures in Space Using the Magnetically Inflated Cable (MIC) System, James Powell, George Maise, John Paniagua, and John Rather, Paper IAC-06-D1.2.09; delivered at the 57th International Astronautical Congress, Valencia, Spain, October 2-5, 2006 [12 pages].

Ibid, viewgraphs of oral presentation at 57th International Astronautical Congress, October 2-5, 2006 [28 pages].

MIC: Magnetically Deployable Structures for Power, Propulsion, Processing, Habitats, and Energy Storage at Manned Lunar Bases, James Powell, George Maise, John Paniagua, and John Rather, to be delivered at STAIF-2007 Conference, Albuquerque, New Mexico, February 11-15, 2007 [9 pages].

Magnetically Inflated Cable (MIC) System for Large Scale Space Structures, James Powell, George Maise, John Paniagua, and John Rather, NIAC (NASA Institute of Advanced Concepts) Phase 1 Report, May 1, 2006 [162 pages].

MIC—Large Scale Magnetically Inflated Cable Structures for Space Power, Propulsion, Communications and Observational Applications, James Powell, George Maise, and John Rather, delivered at SPESIR-2010 International Forum, John Hopkins Applied Physics Laboratory, February 23-26, 2010 (12 pages).

Ibid: Viewgraphs of oral presentation at SPESIF International Forum, February 23-26, 2010 (31 pages).

A Development and Test Program for the Magnetically Inflated Cable (MIC) Large Space Structures System, James Powell, George Maise, and John Rather, delivered at SPESIF-2011 International Forum, Johns Hopkins Applied Physics Laboratory, March 15-17, 2011 (14 pages).

Ibid: Viewgraphs of oral presentation at SPESIF International Forum, March 15-17, 2011 (26 pages).

List of Maglev Reports

1. Powell, J. and Danby, G., 1966. "High Speed Transport By Magnetically Suspended Trains", Paper 66WA/RR-5, ASME Winter Annual Meeting, New York, NY. Also, Powell, J. and Danby, G., 1967 "A 300 mph Magnetically Suspended Train, Mech Eng 89, p. 30-35

2. Powell, J. and Danby, G., 1969. "Electromagnetic Inductive Suspension and Stabilization System For A Ground Vehicle"; US Patent 3,470,828

3. Powell, J. and Danby, G., 1969, "Magnetically Suspended Trains: The Application of Superconductors to High Speed Transport", Cryogenics and Industrial Gases, 4 (10), p.19

4. Powell, J. and Danby, G., (3. 1970. "Dynamically Stable Cryogenic Magnetic. Suspensions for Vehicles in Very High Velocity Transport Systems". Recent Advances in Engineering Science, Gordon and Breach, Vol 5: p. 159-182

5. Powell, J: and Danby, G., 1971. "The Linear Synchronous Motor and High Speed Transport" Proc Intersociety Energy Conversion Eng. Conference, Boston, MA, p. 11 8-131

6. Powell, J. and Danby, G., 1971. "Magnetic Suspension For Levitated Tracked Vehicles" Cryogenics 11: p. 192-204

7. Danby, G. and Powell, J., "The Central Role of Cryogenics in Magnetically Levitated High Speed Trains" Proc. Of XIII International Conference on Refrigeration, Washington, DC

8, Powell, J. and Danby, G., 1971. "Cryogenic Suspension and Propulsion systems for 200 —2000 mph Ground Transport" Proc. Cryogenic Society of America Conf. on Applications of Cryogenic Technology, Vol 4, p. 299-332

9. Danby G. and Powell, J., 1972. "Integrated Systems for Magnetic Suspension and Propulsion of Vehicles" Proc. 1972. Applied Superconductivity Conf., Annapolis, p.120-126

10. Powell, J. and Danby, G., 1972. "Integrated Magnetic Suspension and Propulsion Systems" Proc. IEEE Meeting of Industrial Applications Society. Philadelphia, PA (10 pages)

11. Danby, G. , Jackson, J., and Powell, J. 1974. "Force Calculations for Hybrid (Ferro — Null Flux) Low Drag Systems, IEEE Trans on Magnetic Mag 10, p. 443.446

12. Danby, G. and Powell, .1., 1974. "Hybrid Superconducting Magnetic Suspensions for Very Efficient High Speed Ground Transport" Proc. 1974, Applied Superconductivity Conference

13. Danby, G. and Powell, J., 1988. "Design Approaches and Parameters for Magnetically Levitated Transport Systems" Proc. 2nd Annual Conference on Superconductivity and its Applications, Elsevier Science Publishing

14. Powell, J,. 1992. "Large Scale Implementation of Maglev in the United States" AAAS Symposium on Maglev Transport, Chicago, IL (10 pages)

15. Powell, J. and Danby, G., 1995. "Passenger and Freight Maglev for the US'-, Proc of the Future Transportation Technology Conf., Costa Mesa, CA. SAE Paper 95-1921

16. Powell, J., 1995. "The Application of Maglev Technology to Intermodal Transportation". Proc of National Aviation and Transportation Center, 4th Annual Symposium on Global Intermodalism and Economic Development, July 24-27, 1995

17, Powell, J. and Danby, G., 1996. "Integrating Passenger and Freight Service: Maglev Technology Approach", Proc of 514 Annual Symposium on Intermodal Transportation. Bordeaux, France (41 pages)

18. Powell, J. and Danby, G., 1998. "Transport by Magnetic Levitation", Encyclopedia of Applied Physics, Vol 22, p 233-261

19, Powell, J. and Danby, G., 1989. Co-Chairman, Maglev Technology Advisory Committee for US Senate Committee on Environment and Public Works, "Benefits of Magnetically Levitated Transport for the United States", Volume 1, Executive Summary, (30 pages) and Volume 2, Technical Report (238 pages)

20. Powell, J. and Danby, G., 2000. "Magnetic Levitation: A New Mode of Transport for the 21th Century", Lecture Given at the Award of the 2000 Franklin Medal for Engineering to Powell and Danby by the Franklin Institute (50 Pages)

21. Powell, J. and Danby, G., 2002. "Maglev 2000 'Transportation Technology", Final Report Vol 1 (347 pages) and Vol 2, (452 pages) Federal Railroad Administration and Florida Department of Transportation (849 total pages)

22. "Florida Maglev Deployment Program Kennedy Space Center Circulator National Demonstration Project", Tilden, Lobnitz & Cooper, 2002, (238 Pages)

23. Powell, J. and Danby, G., 2005. "Final Report, Federal Transit Administration, Maglev 2000 Project FL-26-7023", Maglev 2000 of Florida Corporation (100 pages)

24. Powell, J. "Maglev Presentation", ETA Low Speed Urban Maglev Workshop. W.Kulyk, Director, Office of Mobility Innovation, September 8.9, 2005, Washington, DC (30 pages)

25. Powell, J. and Danby, G. "Integration of Maglev Guideways with Railroad Track: The MERRI System", Report DPMT-1, October 5, 1996 (70 Pages)

26. Powell, J. and Danby, G. "Maglev Vehicles", IEEE Potentials, p.7-12, October/November, 1996

27. Powell, J. and Danby, G., "The Development of Maglev-Yamanashi and Beyond", Invited Talk at Dedication of Japan Railways Yamanashi Maglev Test Line, April 4, 1997) (23 Pages)/

28. Powell, J. and Danby, G., "The M-2000 Maglev System for the United States", Presentation to the FRA Maglev Advisory Committee, March 24, 1997 (28 Pages)

29. Powell, J. and Danby, G., "Maglev Technologies for Combined Freight and Passenger Movement — Application to Industrialized and Rapidly Industrializing Nations", 6th International Symposium in Intermodal Transportation, Mexico City June 18-20, 1997 (29 Pages)

30. Powell, J, and Danby, G., and Bragden, C., "Developing A High Speed Maglev Land Bridge in Central America" 66 International Symposium on Intermodal Transportation, Mexico City, June 18-20 1997 (28 pages)

31. Powell, J. and Danby, G., "The Water Train: Long Distance Transport of Water by Maglev", Report DPMT-3, December 13, 1997 (34 pages)

32. Powell, J., "Electrical Power Storage Using Maglev — The Maglev Power Storage System", Report DPMT-14, November 1, 2000 (126 Pages)

33. Powell. J. and Danby, G., "Low Speed Application of Superconducting Maglev", presentation to Transportation Research Board, January 14, 1999, Washington, DC (30 pages)

34. Powell, J., "Cost Projections for the M-2000 Maglev System" Report M-2000/002, presentation to the FRA Workshop, August 24, 1999, Washington, DC (30 pages)

35. Danby, G., "The M-2000 Maglev System" Report 2000/001, Presentation to the FRA Workshop, August 24, 1999, Washington, DC (26 Pages).

36. Danby G., and Powell, J., "Progress in Design and Testing of Maglev 2000 Technology", Report No. M-2001, Presentation to Transportation Research Board 2000 Annual Meeting, Washington, DC, January 9-13, 2000 (29 pages1

37. Powell, and Danby, G., "Maglev: An Evolutionary Technology for the Transport of Freight, People, and Resources", Presented at International Congress on the Implementation Follow Up of Habitat Agenda in Islamic Cities, Tunis, Tunisia, March 24-26, 2000 (46 pages)

38. Powell, J. and Danby, G., Morena, J, Wagner, T, and Smith, C., "The Maglev 2000 Urban Transit Systems", July 31, 2002 (15 Pages).

39. Powell, James, and Danby, Gordon, "The 2nd Generation Maglev 2000 Transport System: Design, Technology, Status, and Future Applications, Maglev 2000 of Florida, Sept. 2006 (166 pages)

40. Jordan, James, and Powell, James " Maglev Transport – A Necessity in the Age of No Oil", presented Capital Science 2008, Arlington, VA., National Science Foundation, March 29-30, 2008 (40 pages)

41. Ibid, PowerPoint presentation, March 29-30, 2008 (25 pages)

42. Powell, James and Gordon Danby, "Energy Efficiency and Economics of Maglev Transport", Presented at 2008 Advanced Energy Conference, Stony Brook University, NY, November 19-20, 2008 (22 pages)

43. Ibid, PowerPoint presentation, November 19-20, 2008 (18 pages)

44. Powell, James, et al, "A National Maglev Network for the US – Design and Capabilities", Presented at Maglev 2008, 20th International Conference on Magnetically Levitated Systems and Linear Drives. San Diego, California, December 15-18, 2008 (9 pages)

45. Danby, Gordon, et al, "Fabrication and Testing of Full-Scale Components For the 2nd Generation Maglev 2000 System, 20th International Conference on Magnetically Levitated Systems and Linear Drives, San Diego, California, December 15-18, 2008 (10 pages)

46. Powell, James, et al, "Adaptation of the LIRR System to Maglev for Faster, More Convenient, and Lower Cost Service", Presented at 2nd Advanced Energy Conference, Stony Brook, Long Island, New York, November 18-19, 2009 (18 pages)

47. Ibid, PowerPoint Presentation, Nov 18-19, 2009 (29 pages)

48. Griffis, F.H. (Bud), et al, "How Maglev Can Enable Stewart Airport to become the 4th Major Airport for the NYC Region". Presented at the 2nd Advanced Energy Conference, Stony Brook, Long Island, New York, November 18-19, 2009 (21 Pages)

49. Ibid, PowerPoint Presentation, Nov 18-19, 2009 (23 Pages)

50. Powell, James, et al, "The West Coast Maglev Network Transport for the 21st Century, Maglev 2000, January 20, 2010 (39 pages)

51. Powell, James, et al, "The New York Maglev Network – A New Transport System for the 21st Century, Maglev 2000, January, 2011 (12 pages)

52. Powell, James, et al, "Maglev Energy Storage and the Grid", Presented at the 2010 Advanced Energy Conference, New York, NY, November 8-9, 2010 (11 pages)

53. Griffis, F.H, (Bud), et al, "Feasibility Study for New Danby Powell Maglev System & Preliminary Route Study (New York City to Stewart International Airport),

Preliminary Guideway Plans and Updated Cost Estimates, Polytechnic Institute of NY, January 2011 (144 pages)

54. Powell, James, et al, "The Maglev America Project: A 29,000 National Maglev Network for the United States", Presented at Maglev 2011, 21st International Conference on Magnetically Levitated Systems and Linear Drives, Daejeon, Korea, Oct 10-13, 2011 (15 pages)

55. Ibid, PowerPoint Presentation, October 10-13, 2011 (17 pages).

56. Griffis, F.H.(Bud), et al, "Adaptation of Existing Railroad Trackage for Levitated Maglev Vehicles", presented at Maglev 2011, 21st International Conference on Magnetically Levitated Systems and Linear Drives, Daejeon, Korea, Oct. 10-13, 2011 (6 pages)

57. Powell, James, et al, "Large Scale Storage of Electrical Energy Using Maglev", presented at Maglev 2011, 21st International Conference on Magnetically Levitated Systems and Linear Drives, Daejeon, Korea, Oct 10-13, 2011 (19 pages)

Maglev 2000 Proprietary Reports

1. Powell, J., "Geometry and Magnetic Forces on the Electromagnetic Loops for the M¬2000 Narrow Beam Guideway", Memo M-2000-JP-3-01-6/14/96, June 14, 1996 (61 pages).

2. Powell, J., "Fringe Fields From SC Magnets", Memo M-2000-JP-3-02-7/26/96, July 26, 1996 (13 pages)

3. Powell, J., "Computer Analysis of Magnetic Field Distributions From M-2000 SC Quadrupole Arrays: Nature and Requirements for Phase A Studies", Memo M-2000-JP¬03-8/03/96 (32 pages).

4. Powell, J., "Refrigeration Systems for High Temperature Superconducting Vehicle Magnets", Memo M-2000-JP-4-01-8/17/96, August 17, 1996 (6 pages).

5. Powell, J., and Maise, G, "Vehicle Design and Mass Budget", Memo M-2000-JP/GM-01-1-8/24/96 August 24, 1996 (14 pages)

6. Maise, G., "Potential Damage to Sides of Vehicles from Windborne Debris", Memo M-2000=GM-1-01-09/26/96 September 26, 1996 (4 pages)

7. Maise, G., "Aerodynamic Lift to Augment Magnetic Levitation", Memo M-2000-GM¬1-03-12/11/96 December 11, 1996 (5 pages)

8. Powell, J., "Optimization of Placement of LSM Winding", Memo M-2000-JP-3-06¬12/31/96, December 31, 1996 (5 pages).

9. Maise, G., "Vehicle Drag Coefficient", Memo M-2000-GM-1-02-10/31/96 October 31, 1996 (9 pages)

10. Maise, G., "Design of M-2000 Vehicle", Memo M-2000-GM-1-04-1/21/97, January 31, 1997 (6 pages).

11. Powell, J., "Vehicle Speeds and Accelerations on Proposed Brevard County Route", Memo M-2000-JP-01-02-1/10/97, January 10, 1997 (11 pages)

12. Maise, G., "Tilting of M-2000 Guideway and/or Vehicles to Eliminate Lateral Forces on Passengers" Memo M-2000-GM-1-05-07/11/97, July 11,1997 (6 pages)

13. M-2000 Program Review; Presentation to C. Smith, Florida Department of Transportation, August 1, 1947 (33 pages)

14. Maise, G., "Aerodynamic Loads on the Guideway Structure Due to Hurricane — Level Wind Forces", Memo M-2000 — GM-1-07/12/31/97, December 31, 1997 (10 pages)

15. Powell, J., "Design of Guideway Loops and Panels, Memo M-2000-JP-3-04¬11/15/96, November 15, 1997 (24 pages)

16. Powell, J., "Reduction of Fringe Fields in the Passenger Cabin by Use of Asymmetric Quadrupoles", Report DPMT-6 December 15, 1997 (7 pages)

17. Powell, J., "The NOVI Ride Control System: A Method for Eliminating Vibration and Maximizing Ride Comfort in Maglev Transportation Systems", Report DPMT-4, December 15, 1997 (23 pages)

18. Powell, J., "On Surface Maglev Guideways", Report DPMT-2, September 15, 1997 (11 pages)

19. Powell, J., "Magnetic Anchoring of Maglev Guideway Beams", Report DPMT-5, December 15, 1997 (22 pages).

20. Powell, J., "Morena, G., Powell, J., and Danby, G., "Transport of Bulk Cargo for Above and Underground Mining by Low-Cost, High-Speed Magnetically Levitated Vehicles", Report M-205, National Mining Associates 21" Annual Transportation and Distribution Seminar, January, 1998 (15 pages)

21. Powell, J., ed., "Cost Projections for the M-2000 Maglev System", Report DPMT-20, May 15, 1999 (192 pages).

22. Powell, J., "The Matrushka Magnet – A Low Cost Ultra-Low Refrigeration Load Magnet System", Report DPMT-9, April 1998 (113 Pages)

23. Powell, J., "Non-conventional Methods for Large Scale Manufacture of Maglev Guideway Loops", Report DPMT-7, February 15, 1999 (31 pages)

24. Powell, J., "Analysis of Eddy Current Heating in Guideway Conductors and Determination of Allowable Limits on Conductor Size", Report DPMT-8, February 15, 1999 (37 pages)

25. Lazareth, O., Skaritka, J., and Powell, J. -- "Comparison of Analytical Calculations of Magnetic Forces for the M-2000 Maglev Systems with Experimental Measurements", Report DPMT-11, August 11, 1999 (35 pages)

26. Powell, J., "Power Transmission and Distribution Architecture for the M-2000 Maglev System" Report DPMT-12, December 15, 1999 (76 pages)

27. Powell, J., "Levitation, Propulsion, Power, and Braking Systems for Maglev 2000 Vehicles", Report DPMT-22, May 18, 2000 (42 pages)

28. Powell, J., "Communications, Control, and Safety for the M-2000 Maglev System', Report DPMT-23, May 18, 2000 (13 pages).

29. Maise, G., "Design of the M-2000 Maglev Passenger Vehicle, Report DPMT-21, May 30, 2000 (27 pages)

30. Powell, J., "The IRT Levitation Demonstration", Report DPMT-26, May 2001 (20 Pages)

31. Skaritka, J., "Superconducting Magnet Design', Report DPMT-24, May 2001 (23 Pages)

32. Harmer, E, Danby, G., Lazareth, O., and Powell, J., "Experimental Measurements of the Magnetic Forces Between the Maglev 2000 Superconducting Quadrupole and Powered Guideway Loops, with Comparison to Values Predicted Using the Maglev 2000 Computer Code", October 15, 2002 (57 pages)

33. Lazareth, O. and Powell, J., "Computer Analyses of Magnetic Forces Between the M-2000 Vehicle and Powered Guideway Loops", Report DPMT-25 (59 pages)

34. Lazareth, O. and Powell, J. "Planar Guideway Performance of Compact Urban M-2000 Revenue Vehicle Part L Levitation and Stability Performance on Non-Powered Guideway", Report M-2000 PG-2-1 (73 pages)

35. Lazareth, O. and Powell, J., "Planar Guideway Performance of Urban/Suburban M¬2000 Maglev Revenue Vehicle, Part 1: Levitation and Stability Performance on Non-Powered Guideway" Report M2000 PG-1-1 (172 pages)

36. Lazareth, O. and Powell, J., "Propulsion and Power Performance of Maglev 2000 Vehicles on Planar Guideway", Report M-2000 PG-3-1, June 1, 2003 (114 pages)

About the Authors

James R. Powell, Ph.D. is a Director of the MAGLEV 2000 of Florida Corporation

Dr. Powell and his colleague, Dr. Gordon Danby, are the recipients of the 2000 Benjamin Franklin Medal in Engineering for their invention of superconducting Maglev transport. The medal was awarded to Drs. Powell and Danby by The Franklin Institute "for their invention of a magnetically-levitated transport system using super conducting magnets and subsequent work in the field." The Franklin Institute awards medals annually in recognition of the recipients' genius and civic spirit and in memory of the Institute's namesake, Benjamin Franklin, who exhibited those same qualities. Some noted past recipients of the Franklin Institute medals include Alexander Graham Bell, Thomas Edison, Neils Bohr, Max Planck, Albert Einstein, and Stephen Hawking.

He was a senior scientist at Brookhaven National Laboratory (BNL) from 1956 through 1996. His experiences have led to significant advances in the design and analysis of advanced reactor systems, cryogenic and super-conducting power transmission, plasma physics, mine safety, fusion reactor technology, electronuclear (accelerator) breeder systems, transmutation of nuclear wastes, space nuclear thermal propulsion, electromagnetic hypervelocity guns, hydrogen and synthetic fuels, and transportation infrastructure.

He holds patents for the Particle Bed Reactor (PBR) for nuclear rocket propulsion, the use of aluminum structure in fusion reactors; blankets employing solid lithium ceramics and alloys for tritium breeding; and, demountable super conducting magnet systems and the Advanced Vitrification System (AVS) for high-level nuclear and toxic wastes. He and Dr. Danby are the holders of the first patent for superconducting Maglev in 1968, as well as many recent patents on their 2nd generation advanced Maglev system.

Dr. Powell holds a Bachelor of Science in Chemical Engineering from the Carnegie Institute of Technology and a Doctor of Science in Nuclear Engineering earned in 1958 from the Massachusetts Institute of Technology. Dr. Powell has published almost 500 professional papers and reports. He is a member of the American Nuclear Society.

Physical Society. In 1983, the New York Academy of Sciences honored Danby with the Boris Pregel Award for Applied Science and Technology.

Dr. Jesse Powell, Ph.D. is Founder and President of Maglev Strategies, in which capacity he works to identify new markets and opportunities for maglev technologies. He coordinates between Maglev 2000, Inc. and third-party companies in the scoping of new projects, and manages technology transfer issues. Currently, he is focused on maglev space launch, maglev energy storage, and maglev water transport as the areas most likely to attract funding in the United States.

Jesse Powell at Brookhaven National Lab on the Occasion of the 50th Anniversary of the invention of Superconducting Maglev by Gordon Danby and his father, James Powell

https://www.bnl.gov/video/index.php?v=514

From 2002 to 2013, Dr. Powell has worked in the field of Oceanography. He worked at Scripps Institution of Oceanography, where he studied the impact of ocean fronts and mesoscale ocean structures on plankton distributions. During this time, he also worked on a range of technology projects spanning from mission designs for the **exploration of Mars and Europa, to the use of autonomous underwater vehicles to map ocean life, to** machine vision systems for plankton identification and the automatic classification of fish eggs of important species for habitat mapping.

Dr. Powell holds a BS in Biology and BA in French Literature from University of California, San Diego, a MS in Molecular Biology from San Diego State University, and a PhD from Scripps Institution of Oceanography.

James Jordan is the founder and President of the Interstate Maglev Project and Executive Vice President of Maglev 2000, a co-author with Powell and Danby of "*The Fight For Maglev*", co-author with James and Jesse Powell of *"Silent Earth, Will Humans Give Up Fossil Fuels"*. He is also the managing editor of "*Maglev America*" and "*7 Projects for a Better World*". He can be reached at

james.jordan@cox.net

The energy crises of the 1970s focused the Navy career of James Jordan. The era of scarce oil and rapid increases in oil prices dramatically introduced Commander Jordan to the national military and economic security consequences of America's growing dependence on oil. Commander Jordan served as the director of the Navy Energy R&D program office in the Pentagon. As director, he developed strategies and technologies aimed at sustaining military and national economic security in the new oil reality.

In 1979, Mr. Jordan retired from the Navy and became a senior policy advisor to the late Senator John C. Stennis, Chairman, Armed Services Committee and Defense Appropriations Committee. In this capacity, Mr. Jordan was a Senate staff leader in energy, transportation, environment, and agricultural policy.

In 1988, after leaving the U.S. Senate, Mr. Jordan founded several entrepreneurial ventures directed toward development of environmentally sustainable energy and economic growth: efficient all-electric Maglev (**mag**netic **lev**itation) transportation, carbon capture and storage, nuclear waste isolation, hydrogen and electric power co-generation, advanced nuclear power generation and earth science data management.[1]

Education: MBA, Harvard Business School, Cambridge, MA; Distinguished Graduate, Industrial College of the Armed Forces at the National Defense University, Washington, DC; BA, University of North Carolina, Chapel Hill, NC, Student Body President and Graduate of Senior High School, Greensboro, NC.

Consortium International Earth Science Information Network, (www.ciesin.org), now located at Columbia University. In 1992, Mr. Jordan introduced CIESIN to the U.N. Conference on Environmental Development (UNCED) in Rio.

www.ingramcontent.com/pod-product-compliance
Lightning Source LLC
Chambersburg PA
CBHW062215220526
45471CB00009B/3204